Cultural Shock

L. Mason Jones served a number of years in the military, and travelling on a so-called 'government service' passport, found himself in such places as south Yemen, Bahrain the Gulf of Oman, Cyprus and Germany. After leaving the services, he became part of the team producing the highly successful business jet, The Hawker 125. He functioned as a quality-engineering inspector with, initially, British Aerospace then Corporate Jets Inc and finally Raytheon USA, the latter of which purchased the thriving business and moved production to the USA. Mr Jones then left the business to concentrate on writing projects. He has three adult offspring and resides in Chester.

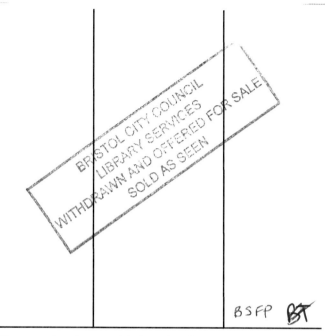

BSFP BT

Please return/renew this item by the last date shown
on this label, or on your self-service receipt.

To renew this item, visit **www.librarieswest.org.uk**
or contact your library

Your borrower number and PIN are required.

LibrariesWest

Cultural Shock

L. Mason Jones

Arena Books

First published in 2020 by Arena Books

Arena Books

6 Southgate Green
Bury St. Edmunds
IP33 2BL

www.arenabooks.co.uk

Distributed in America by Ingram International, One Ingram
Blvd., P.O. Box 3006, La Vergne, TN 37086-1985, USA.

L. Mason Jones

Cultural Shock

British Library cataloguing in Publication Data. A Catalogue record
for this book is available from the British Library.

ISBN-13   978-1-911593-80-5

BIC classifications:- PDA, PGK, RBC, RBGD, PDZ, ABGF, PGS, RNR, TTD.

Cover design
By Jason Anscomb

Typeset in
Times New Roman

INTRODUCTION

C ould it really happen? How many times have we sat, perhaps with a drink or maybe an ice cream in our hand and watched from the comfort of our armchairs or from a cinema seat, the dramatic arrival of alien beings into earth space? Later in the evening, we go off to bed and forget all about it, but how would we handle it in reality? Eventually, we would have to go off to bed, but getting to sleep would be a different matter, and we certainly could not forget all about it. There would be immense shock, panic and social upheaval and we may never adjust to it. We, as humans, are used to having dominion over all other creatures on Earth. How could we handle creatures arriving on Earth that felt they had dominion over us? Have all those movies and television films taught us anything at all in how to become prepared? How is it possible to know how to become prepared?

Whether one believes it or not, it is certainly hard not to be interested in the enormous amount of convincing, but circumstantial evidence and data, including believable abduction cases that seem to indicate that an alien intelligence is already here in our earth space. We find the postulations interesting but we do not really believe them. It is not something we could handle at all.

However, there is one thing we can be fairly certain of and that is, if it is true, then 'they' are not going to simply keep us guessing by a progressive series of encounters and abductions indefinitely. They would have a strategy, a plan, a purpose. There would surely have to be a culmination to it all and, for all we know, the decision on where and when this culmination will take place may have already been made. Humans will soon be on the 'springboard' ready for another 'giant leap' for mankind, Mars and beyond. It may seem a better idea to confront a higher alien intelligence here in our own world, rather than in deep space. In any case, could humanity, who cannot seem to stop killing each other, be viewed as fit and ready to enter the domain, which 'they' may be familiar with, but we are most certainly not?

If there really is a cosmic intelligence operating in earth space, perhaps even from biblical times (as all those encounters with 'angels' in the Bible seem to suggest), many factors indicate that they are seriously advanced and intelligent.

Is Planet Earth the only populated world? It is fairly logical to assume earth-like planets existing with lightning bolts stoking their primordial seas at a similar time to when the process was going on here on Earth or far ahead of the event here as we are comparatively new members on the galactic scene and it is quite likely to be the senior members who are now allegedly looking us over so patiently in our skies. Can so many reports be wrong? Decades ago, there were around 50,000 incidences on record, probably now double that number. When Dr. J Allen Hynek headed the Centre for UFO Studies, the computerised data bank at the Centre had that many reports from worldwide sources. To be sure, the ability of great 'patience' would certainly be required of them if indeed they do represent a controlled intelligent phenomena, since the reports, from the biblical characters to modern day sightings imply such a lengthy period of time. Clearly, in olden times there was no mechanism for logging and recording reports and the true, but obviously unknown number of sightings and events since the days of 'Ezekiel's vision' must be truly enormous. It would appear that certain people have researched old documents and a surprising number of things the people of those times could not adequately explain do crop up.

With regard to some older observations, the Japanese explained a case in 1235 AD that seemed to be UFO activity as "it is only the wind making the stars sway". It seems that a certain General Yoritsume, on a campaign, observed light sources swinging and circling in the south west until the early morning. After a full enquiry, his consultants came up with the above report. Over 700 years later, a very similar report came in from the pilot and crew of an aircraft flying over French Morocco in 1953 over Nouasseur airbase.

A couple of hundred years earlier, Christopher Columbus was reporting similar things seen while patrolling the deck of the Santa Maria with Pedro Guterezz. A "glimmering light, at a great distance; and coming by in sudden and passing gleams". Reports can be extracted from almost any period in written history.

In 1880, Monsieur M.A. Trecul of the French Academy saw a golden-white cigar-shaped object from which a small circular object detached itself. Is this the often reported 'mother ship' and smaller planetary observation craft we hear about?

The author/zoologist, Bjorn Kurten, stated in *Not from the Apes* (Victor Gollancz) that, "It is essential to know where we came from in order to know where we are going". It would be nice but not essential to know where we came from, but we do know where we are going – into the cosmos – there is nothing more certain. The driving quest within us to do so is relentless. When we do achieve this aim, will we simply just be going home?

The human brain has been bestowed with a capacity for mathematics; this is why it will happen, without mathematics it could never happen.

If we include the Moon, human built artefacts, due to this driving quest have now settled down upon three other worlds in the solar system. These artefacts are simply an extension of our own probing exploring, examining and immensely enquiring brain, an organ with amazing capabilities for creativity and advancement that we seem to take so much for granted. Now, humankind has reached out of our solar system altogether in the form of deep space probes, one with a gold-plated plaque revealing to any enquiring cosmic beings who could encounter it, where the probe came from and what the beings looked like that dispatched it. Some people might view the assumption that any beings encountering it will *all* be friendly, as a dangerous exercise, but it is too late to worry about that now. However, we can for the moment, feel quite safe about it, because on its present course it will take so long to reach a star that we will have plenty of time to develop near light speed power sources and go and get it back. Of course, the possibility exists that some exploring alien intelligence might encounter it already on their way to Earth, since we have already told any possible intelligence up to ninety light years away that we do exist, due to our earthly radio and television broadcasts, not to mention the purposeful emissions from the S.E.T.I. programme.

We have already looked over, photographed and deeply analysed our planetary neighbours for any life signs and seem to have rejected them completely, except one. The planet Mars. Of course, there are theories that strange floating life forms may exist in the more temperate layers between the higher and lower extremes on Jupiter, or perhaps under the shifting ice and warmer zones within Europa, an interesting moon of Jupiter, but for surface living, air-breathing life forms, Venus and Mars were initially our hopeful alternative to Earth. With the benefit of hindsight, it would have been amazing for us to have found evidence of any life signs on Venus now, but not when we consider that we could be converting Earth to a Venus-like environment, ourselves, but Mars is still intriguing. We seem to sense that in spite of all the negative reports from our probes, landers and soil analysis, that something profound will, one day, be found there. Before the Vikings landed soil scoop up and analysis, that produced such inconclusive results, it was commonplace for amateur astronomers and science correspondents to be pronouncing on the television that very basic life forms could be found there, such as lichen and mosses and primitive, but multi-cellular life, but after the probes and landers, their postulations changed dramatically. "Nothing could have lived here" or "It's more like the Moon than the Earth", and "Mars is a barren dead world".

At first, the scene before the camera on the Viking Lander sitting on the Martian surface, did quite resemble an earthly desert scene, but the essential difference was that no life form of any description buzzed, crawled, hopped or slithered past the watching eye of the camera that would have been apparent to any alien probe landing in even the most hostile regions on Earth. Nevertheless, water once existed there; the features are sharp and clearly not eroded by wind and weather, as for their age?

Billions rather than millions was the figure mentioned, but given that Martian probes from Earth have had to spend many orbits in the 'wait' mode due to the enormous dust storms, before they could begin their photographing and mapping when conditions cleared, and given that the spreading dark patches in the Martian summer (initially thought to be vegetation from polar melting ice) were, in fact, dust storms, then an enormous amount of surface erosion would have been taking place and this would have logically obliterated those water flow signs long ago, so they cannot surely, be that old. In theory, with so much erosion going on, we should not be seeing any sharp features at all such as faces and pyramids, and what appear to be purposeful structures.

Although the Martian volcanoes are, or were, of the liquid lava type, the primary example of them, particularly Olympus Mons, must have pumped out an enormous amount of dust over their active period. Four huge examples of the larger variety straddle the Martian equator, and the towering example Olympus Mons, is seventeen miles high, or three times the height of Mount Everest. Clear volcanic traces exist in the Martian 'soil' such as iron, Mars is a 'rusty' world.

Picture an alien interstellar probe coming into our solar system and releasing a robotic investigative lander, settling and scooping and analysing the basic 'soil' of the rocky worlds. If one had landed by Mount St Helens in the USA after the eruption, among all that dust would it conclude that life flourished on Earth in abundance when sniffing at the sterile material? This may have been what the Viking Lander scoop up operation was up against with regard to the Martian 'soil', which seemed more like iron oxide than something that would see a green plant flourishing in it. Copious amounts of volcanic dust must be blowing about on Mars. What will we find on Mars when we eventually land, set up a base and really begin to excavate and examine the deeper material of that world?

The answers may astound us, particularly in view of the fact that methane emissions, a signature for the existence of organic life have been detected on Mars, giving rise to the speculation that life may exist below the surface.

If these prerequisites for the possible existence of flowing water were at one time in existence on Mars, then conditions in the habitability zone of our solar system may have been greatly different to what they are today. Mars may have been a far more hospitable world compared to its rather hostile conditions of the present day. What happened long in the past in the zone between the Asteroid Belt and the planet Mars? A planet should have (and may have) existed in that region of tumbling rocks. Every grain of dust and every stone that streaks into our atmosphere as a meteor, including all the asteroids, may have at one time been part of a whole body we could call Planet X.

If we look at any meteorite found on Earth, whether iron-based or carbonaceous chondrite types, they are all dense rocks and must have been subjected to the same process that formed the earthly rocks, i.e. immense planetary gravitational pressure. What process of physics could see their formation coming about from the negligible gravity of small, wispy gas clouds? Surely, they must all have been part of the larger body, or planet, yet we continually hear the expression "It's material leftover from the formation of the solar system". Surely, the sun and the planets would have grabbed all the heavy elements and material and only dust and gas should have remained. In any case, geologists have concluded that they must have once been part of the 'parent body'.

There are many mysteries in the cosmos we have yet to solve. Is Mars 'off limits' to close scrutiny? After the inconclusive Viking Lander, soil analysis programme, which may have been as previously said, simply looking at sterile volcanic dust, humans did not go away and simply give up like we seem to have done in our lunar programme with plans for Moon bases and observatories.

Advanced Mars landers have trundled about finding many interesting features and are continuing to do so, we have not given up with regard to Mars, we continue to peer and probe at that mystifying world and now plans for bases and observations exist for Mars. Will an alien presence be found there also? In 1988, Russia sent two probes to Mars named after one of the Martian Moons. They were known as Phobos I and Phobos II. Phobos I suddenly and inexplicably disappeared on its approach to Mars. Phobos II made it into orbit around that world and began to transmit photographs in the infrared, possibly for underground heat sources. An amazing grid-like pattern appeared, suggesting a symmetrical intelligently constructed underground zone betraying its position in the infrared. If that was not enough, when Phobos II was redirected to photograph the Martian moons it failed, and the last Phobos II transmission showed what appeared to be a 'bright light' approaching the craft. In November 1996, another Russian Mars probe fell into the Pacific after a booster failed to function. Obviously, the Russians were going for another look. Now the NASA 'Pathfinder', with a Mars Rover vehicle, although landing safely seems to have encountered trouble from the word go. However, it did eventually trundle off its track after having emerged from a Balloon type of cocoon. When the euphoria died down it began the serious business of looking for life signs, anything significant will be well digested first by the scientist and decisions will *then* be made on whether the *rest* of us are to be told about it. The Mars rover appears to be having a great success and we have to admire the landing process with parachutes, cables and rockets.

It has been suggested by some writers that the earthly space research programme may have been purposely slowed down with all those early failures on the launch pad up to the narrowly averted disaster of Apollo 13. There have been curious things observed on Mars. The Moon is not the only place where T.L.P., or trans lunar phenomena has been observed. Mars has had its share of what we might term T.M.P. when we speculate about the E.T. 'end game' after their lengthy presence many appear to have a sense of foreboding that something 'is in the air'.

The other creatures of the Earth seem to sense it and draw closer to humans. Ravens move into the cities. Foxes, once rarely seen, now move close enough to humans in some areas in broad daylight, to the point of being hand fed. There was a time when the sight of a squirrel caused great excitement. Now, even they eat out of our hands. Just what is coming?

It would be rather naïve of us to expect, when postulating on the millions of extra terrestrial intelligence that science now accepts must exist, that they will all be tranquil, benign, peaceful-loving entities, all so mentally mature as to have left war and aggression far behind them. A huge portion of the total Earthly landmasses (which was even larger before certain events in Eastern Europe) is home to a regime that totally oppresses democracy and rules in a totally autocratic and totalitarian manner. Many philosophers and prophets of doom have foreseen a world dominated and ruled by such a regime. It is entirely possible that other worlds exist where such a situation prevails and could quite easily reside in an area of space that has been reached by intelligent signals from Earth and now knows we are here.

Warnings have been given (and all ignored) regarding our cavalier attitude to our purposeful signals to other possible planetary systems based on scenarios of past Earth history in the subjugation of other races, not always by military might, but by simple introduction of an advanced culture and its effects on the less advanced one. Old ways, methods, codes of behaviour, languages and traditions have entirely disappeared with the desire for 'modernisation', either freely adopted or imposed on the less advanced race by the more advanced culture, not to mention diseases unfamiliar to conquered races inadvertently introduced to them during the age of discovery.

The strong motivation of commercial gain, and to be fair, to increase safety and survival chances at sea by rescue, saw the development of radio communication (and later television broadcasting) soar ahead at an enormous pace, with no inclination to discuss the departure of such ensuing radiations from planet Earth.

Whether we live to regret it or not, planet Earth has announced its presence in the galaxy and a large portion of the galaxy experiences earthly signals, the fact is that a huge ever-spreading sphere of intelligent radiations is expanding outward from Earth and is now wafting over possible planets with intelligent life forms, 90 light years away from all points on Earth.

It is too late to change our minds or have second thoughts about this fact. All we can do is to hope, just as we do with our purposeful emissions in our search for extra terrestrial intelligences (or S.E.T.I.) programmes that we do not invite any possible 'off Earth' intelligence with evil intent, or a desperate need for another home to our portion of the galaxy.

The scenes depicted in classical science fiction stories may be entirely possible. Beings of another world may exist that have caused irreversible damage to their eco-systems and may have seriously depleted their planetary resources to boot.

If they have recently developed interstellar travel, perhaps with fission powered propulsion systems and have been wondering which way to send their fleet of star ships to extinction or survival, then suddenly, after so much fruitless searching of the heavens for intelligent radiations, they begin to detect our broadcasts, with highly advanced apparatus, able to sift the weaker emissions from the background hiss left over from the colossal event of 15-20 billion years ago. Then this would be an open invitation to them to set course for Earth. They would have nothing to lose and everything to gain.

Such hypothetical entities may not reside at the current outer limits of our radiation, but much closer to home. There are sun-like stars just over ten light years away, where multi-cellular intelligent life forms may have developed. If their science has developed enough to enable them to have prevented leakage of their radiations, but also provided them with sophisticated detection methods, it is not surprising we have not heard a peep out of them.

However, they could have been listening to the sounds of Earth for many years. Put the case that our initial radio emissions reached such a system ten light years away in the early 1930's then allowing for analysis and location, then their departure accelerating to a large portion of the speed of light 'they' could have arrived in our system around 1947. Naturally, the US, having built and delivered nuclear weapons was quite concerned about other nations possessing them but nuclear tests began that could attract E.T attention. 1947 is a very interesting year that witnessed the term 'flying saucer' being introduced to our vocabulary, and the still unsolved and hotly debated 'Rosswell incident', where alien body forms and 'modus operandi' were allegedly whisked away to some remote Hangar 18 facility in the USA after army crash retrieval operations, and has been the subject of an alleged cover-up by ufologists everywhere and hotly denied by the debunkers in the various governmental departments ever since.

To explain the wreckage, the US Government stated it was the result of a crashed secret Balloon Operation called Project Mogul. Since the US had tested and used nuclear weapons in anger. They wanted to know if Russia and other Eastern Nations had them, and 'Mogul', a high altitude detector balloon, was intended to 'listen' for other nations nuclear testing, but we would have to see the debris to decide. With regard to the USA, there is no longer any need for them to deny any cover-up. We should be asking what they have that needs to be so desperately retained. As the documents released under the Freedom of Information Act showed so many deletions but obviously not on the retained copies.

CONTENTS

Chapter I

'VACANCIES' HOTEL EARTH

In spite of all our self-induced propaganda and preparation by hypothesising on the existence of extra-terrestrial intelligence in Earth space, their appearance would be utterly fantastic and traumatic. We talk freely about extra-terrestrials and unidentified flying objects and ancient astronauts. We compile and produce convincing science fiction films and speculate on all the Earth-like worlds that must exist, yet for all this, the appearance in numbers of actual alien beings from another world in Earth space would cause such tremendous social disorientation and cultural shock, that some would not be able to mentally handle it at all and may go completely out of their minds. If the US scientists are aware that E.T. does exist in our Earth space, there would bound to have been discussions regarding E.T.s final plan.

It is for certain that responsible world leaders have seriously considered these factors. Regardless of whether there is any truth in the allegations of their withholding clear evidence of extra-terrestrial presence in Earth space or not, but what administration would want the parcel to stop with them, they are only in power for a few years unless re-elected, they would prefer to pass the parcel to the next administration.

It is also for certain that the reaction from Earth's races would, (although being obviously one of trauma), be far less dramatic, regarding revelations from world leaders regarding alien presence, than it would be from witnessing an unannounced arrival of alien interstellar craft descending over all the capital cities on Earth, literally 'out of the blue', as depicted in all those science fiction movies, and addressing us through controlling the TV channels.

It is perfectly possible that the Earth may be on the threshold of the greatest revelation that human beings have experienced for some 2,000 years.

Therefore, if the various world security agencies, such as the CIA, the FBI, the MOD, or MI5 or any other organisation are, as alleged, in possession of all those secret files on such phenomena, it would be far better to release it all to the masses now, by way of at least some form of preparation, than waiting for such an event and saying "We could have told you so". If they are not in possession of such data, as they continually state, then how is it possible to know how to become prepared, as the threat still remains?

Any country with well-equipped and responsible armed forces, efficient and answerable to their political leaders would not be responsible, if they did not formulate and regularly discuss certain contingency plans for every possible eventuality they could envisage as likely to occur with regard to defence matters, procedures and likely reactions to any hypothetical threat against their country.

After the Argentinean invasion of the Falkland Islands for example, the quite massive military and logistics operation of sending equipping and back-up supplies, fighting troops some 800 miles into the South Atlantic, did not all come about due entirely to a hastily convened cabinet meeting. The threat was known about and planned for in advance, that is what contingency planning is all about and involving massive logistic planning.

Clearly, a file existed and within its enclosures, were all the details, plainly spelled out on such matters as the regiments who would be involved, the equipment, weaponry, ships, commanders and commandeering authority for civilian vessels, landing sites and so forth. All the information in the file would have been compiled, however, hypothetically and loosely formulated, in reaction to intelligence received and gathered, about the possible enemy or aggressor, his strengths, weaknesses, objectives, possible occupation zones, weapon technology and armed forces potential.

Do such files exist in the defence departments of Earth's nations, with regard to any possible threat or arrival of beings not of this Earth? How would one title such a file? 'Possible' E.T. Appearance in Earth space. Actions and options?

How would they raise the first enclosure? Where would the aforesaid intelligence come from? Clearly, in this event, it would be from the hard core and most convincing encounters of alleged E.T. entities of the first, second and third kinds as classified by the late Dr J Allen Hynek. This gentleman, although formerly a sceptic and working as a so-called 'de-bunker' of such phenomena, became convinced later in life of the distinct possibility (no doubt due to the very large amount of convincing cases) of their actual reality. He then became worrisome to his bosses; he was only supposed to scientifically debunk cases re:- UFO's

There would, of necessity, have to be a thesis of possible intent, combined from reports on alleged activity and interest on earthly installations. If such interest was predominantly in regard to defence establishments, that would be quite significant. If the interest was largely in Earth's fauna, soil samples, minerals, water reserves and so forth, this may only indicate an ecological interest in a life supporting planet and perhaps indicative of a passive interest only. Where the problem arises in regard to the mass of reports on seemingly alien craft and their occupants, interest seems to have been shown in every aspect of Earth, coupled with a distinct reluctance to contact or be observed by

"Earthlings", and a distinct interest in nuclear power plants and nuclear weapon storage facilities.

With regard to alien hardware, although in the main, loosely conforming to the circular form, every shape from bullet, beehive to pancake form has been reported, and investigators of UFO encounters have up to 60 distinct shapes on file, which help them to classify the various reports. Many modern sightings report huge triangular shapes, which many believe are highly advanced craft 'developed in black projects' by the US. Area 51 'watchers' using very sophisticated high resolution photographic techniques have noted an additional massive recently constructed hanger type building to house such craft.

Many files (although with obliteration of the text in many cases) have been released due to pressure from legislation on public access. Many have been retained and these could have a security classification enabling them to be retained 'for reasons of national security' for anything up to 100 years by Government agencies. Among these cases would be those reported by mature and reliable witnesses of good standing and by naval, military and air force personnel, particularly by pilots, both service and civilian.

There would be cases backed up by not only ground observers but also air and ground radar sighting and by aircrews despatched to investigate such phenomena.

No doubt, much film footage exists taken by pilots whose aircraft radar sets and gun cameras have locked onto unknown objects weaving away and clearly taking precautions and evasive action under intelligent control to evade identification. These would be classified as extremely convincing encounters of unidentified flying objects, bearing in mind throughout, of course, that 'unidentified' does not necessarily mean 'extra-terrestrial', but it would be worrisome if they belong to an earthly power or potential enemy.

However, so many cases exist of such craft, carrying out these violent and seemingly evasive manoeuvres as to make their earthly origin unlikely in the extreme. In some cases, the objects are not only lost off the observer's radar screens but also are seen to dissolve or disappear, by independent observers, into thin air (or a possible different dimension), perhaps indicating dematerialisation. Reliable reports of this type would remain 'under wraps' for a very long time and be protected from becoming public knowledge by the aforesaid reasons of national security, and no amount of pressure from Freedom of Information laws would secure their release. Their ability to defend would not be enhanced by their retention, unless the 'retainers' could learn something and perhaps duplicate their seemingly incredible power sources and (as we are continually assured) no 'nuts and bolts' extra terrestrial craft exist in captivity ... there is little chance of any defence organisation *ever* learning anything about the said power sources, so they may as well release the information if for no other reason, than a contribution to the public awareness and preparation plan.

Of course, the reason such data is not released is simply, that along with it, would have to be an admission that the defence organisations do not know how to deal with it, and know little, if anything, what the root cause of it is, and this would not enhance their reputation as being capable of properly defending their lands, and that in its entirety, is the sole reason for the retention of such data. One would imagine that the alleged E.T.'s having observed and interacted with humankind for so long would by now be well aware of humanity would react to their appearance.

So then, are these type of files and reports from which data would be drawn to formulate some kind of contingency plan? To be sure, if alien entities do arrive en-masse in Earth space, it would be, as I said in the Introduction, a 'second coming'. Their intelligence gathering, studies and observations would have gone on for some time prior to their decision to appear. With our organised S.E.T.I. programmes and our seemingly urgent need to contact other world beings by shouting out our position in the galaxy, and our cavalier attitude to our expanding radiations, have we seriously considered for long enough, the possibility of attracting alien entities with evil or aggressive intent? Perhaps more than one species? Are we inviting another 'war in heaven' but instead of reading about it in the Bible, we witness it for ourselves with different alien species fighting for domination of the Earth and its resources?

However, even if the alleged E.T.'s are peaceful and benign, the inescapable fact is there must be an 'end game' or final conclusion to their operations but then we could ask 'would a final appearance to humanity be essential?' They could simply just go away. The toughest decision they would ever have to make, regarding human reactions and current behaviour patterns would be when or how to go about a conclusion of their operations, and they would be well aware, humans would not approach them in a subservient manner bearing gifts. So, why not just go away and perhaps return when humanity had matured somewhat towards our mental attitude.

The odds in support of many other planets orbiting other stars must be enormous and once accepting that, we can look at the calculation of the possible 75 billion sun-like stars said to exist throughout our galaxy alone. At least some of them should have systems with perhaps one earth-like planet nestling comfortably within that star's habitability zone with conditions existing on its surface favourable for multi-cellular life to have evolved. Perhaps the seemingly adequate humanoid type shape with bipedal locomotion and manual dexterity being apparent in its life forms and enjoying a good advanced level of technology.

However, since the galaxy is 100,000 light years from end to end, we need not concern ourselves about many of them unless we do circumvent the Einsteinian laws with power sources still within the realms of speculation or science fiction, and though we may have distant neighbours we also have a few much closer to home, and we will look at these in Chapter Two.

And so, the final act of an alien plan that culminates in their eventually revealing themselves to humankind, would have been preceded by enormous and intense intelligence gathering of information of all descriptions about planet Earth, atmospheric analysis and bacteriological content, makeup and proportion of atmospheric gasses, analysis of fauna, vegetation, mineral contents of soil, sea water analysis and depth, marine life, geological and seismic activity, continental drifts and tectonic plate pressure zones, all the volatile areas and all of the reservoirs, utilities and freshwater supplies. The insects and birds of the air and finally earthly creatures of all descriptions, not least the human entity.

To accomplish all this would necessitate organised and well-equipped bases to work from, in the remote areas possible.

Of course, some view the UFO's as 'ancient', we cannot assume the whole rigmarole began in the mid forties, and to be sure, cases of strange aerial craft and UFO activity seem to permeate all the generations since our recorded history began, and this would not lend itself to the assumption that aliens with Earth domination and evil intent are present in our skies, but rather the opposite. Cool, benign intellects, carefully laying their plans against (or for) us. It has been suggested that humanity is the result of some 'alien' experiment'. If so, there would be a responsibility automatically handed down to their descendants that have observed us for centuries. Generations of their kind could have been patiently studying Earth and any final earthly revelation of their life forms to us would be quite profound, as it would be the final culmination and implementation of those plans. What are we to make of all those legends of 'Sky people', 'gods, Initiators, depictions of celestial orbs and spheres and people in worshipful poses, kneeling, looking up at them– what were they seeing?

In the short time that Earth's peoples have been recording their history, it is clear we once thought the Sun and stars all revolved around us, such importance did we attach to ourselves. We now estimate some 95 billion sun-like stars to exist in our galaxy alone and possibly some 300 million planets on which a technical civilisation could well have arisen, and the likelihood of twice that number bearing basic but multi-cellular life forms. Formidable numbers indeed, and numbers clearly indicating a distinct lack of importance of the human life forms in the cosmic scheme of things, but they are only estimates and may be totally wrong. However, for all that we still could be alone in this vast cosmic conglomeration of chaos and order hand in hand, lonely voices shouting into the void and with nobody listening. To be sure, with regard to all the indications from the S.E.T.I. and sky search programmes scanning all these frequencies for at least forty years, nobody, it would appear, is listening, and if they are, they do not seem to be talking. However, what if they have been listening. We have shouted out to the universe loud and clear that 'we are here'. Hotel Earth is open; the lights are on, the 'vacancies' sign is lit up. Can we really be surprised if the guests start to arrive?

The fact that any extra terrestrial beings may have arrived in our system in the first place clearly implies greater and more advanced technical capability and quite obviously they would have to be confident of being able to correctly compute and deal with any adverse reactions from the planet's population, so it must be assumed that the reports of E.T. craft with the occupants emerging from the craft and going about their business, implies little fear (in spite of their rather quick departures often reported) of any human adverse reactions.

Therefore,, bearing all this in mind, there would seem to be little point in retaliation of any description. However, of course, that is not the human way. We like to feel we do not throw in the towel too easily. There almost certainly would be an attempt at retaliation, for we would naturally assume (due entirely to our own actions through Earth's history), a wish to dominate and control us. No matter how convincing the alien message may be that their purpose is to help and assist humanity, it would always be suspect.   Any thoughts of the possibility of the E.T. presence wishing only to help, guide or direct us for our own ultimate benefit, may not enter our heads.   If all the circumstantial evidence of an E.T. presence in Earth space is wrong and all we are seeing in our skies is misunderstood natural or meteorological phenomena the physicists cannot explain, all well and good, but try and convince an abduction victim of this.

There is still time for the Think Tank, the 'Majestic 12', or a chosen body of physicists, scientists, exobiologists and their learned ones to formulate a policy and recommend a course of reaction to some future hypothetical 'coming' as our radiations approach other worlds. If an alien race has been observing Earth and its peoples for a long period of time, a good proportion of that time would have been dedicated to the Perplexing human behaviour, and if E.T. has been involved with humanity as long as seems to be indicated in human affairs and history, they ought to be well aware of exactly how we would react, this may explain their reticence or for all their assumed intellect) 'inability' to formulate a policy on just how to go about their REVELATIONS.

Many of the convincing reports of the late fifties and early sixties filed by police officers and other reliable witnesses, such as the Lonnie Zamora case (Socorro, New Mexico), may have had more 'down to earth' connections.

Research and trials on vertical takeoff and landing vehicles, in preparation for the ultimate Moon landings (not too far away), would have been taking place in those remote areas. Most certainly New Mexico is an area long associated with UFO activity such as the famous 'Rosswell Incident', but also this area was (and is) the scene of much flight testing of classified military aircraft in the experimental stages.

The Lonnie Zamora case took place just five years before the Apollo II lunar landing, and a vertical takeoff and landing vehicle was sighted by Officer Zamora in Socorro, New Mexico in 1964. He witnessed a white or silvery object on four legs from the brow of the hill, which he originally thought was an upturned automobile. Two figures were standing near it in white 'suits' or overalls who entered the craft and it took off vertically and disappeared. The case was well investigated and there seemed to be no inconsistencies or assumptions that Officer Zamora was hallucinating or fabricating the story, and in any case his report was reinforced by evidence on the ground, depression burnt brushwood etc.

Apparently, this case has never been satisfactorily explained and if it was a lunar excursion module development test, surely after all this time and obsolescence of the equipment, the authorities could make it plain whether it was any of their hardware that was witnessed. The time and date are quite precise as April 24th 1964 at 6.00pm. If the unexplained earlier cases were admitted to by NASA or the military, it would clear up many 'COLD CASES' and also the speculation and wonderment regarding the phenomena.

An interesting footnote to this sighting is the special marking on the craft that was mentioned by the police officer. The same marking had been noticed on the Moon and I have related these details in *When the Moon Came*.

It is quite a simple matter to hurl a pie pan or a car hubcap of the old-fashioned variety, photograph it against the sky, and get a statement that the negative had not been tampered with. It would be almost impossible to ascertain velocity or whether it was a small object nearby, or a large object far away, and just such an object was photographed by a Mr Paul Trent on May 11th 1950 at McInnville, Oregon, USA. It was very 'pie pan' shaped, but had a central protrusion. However, four years later in 1954, an object exactly the same in appearance was photographed by a pilot flying over Rouen in France. Furthermore, the McInnville sighting was case No. 46 in the Condon Report and was one of the few UFO reports in which all factors investigated geometric, psychological and physical appeared to be consistent that an extraordinary flying object, silvery metallic, disc shaped and tens of meters in diameter was witnessed.

On more than one occasion, there has been consistency in certain factors of UFO reports from very widely separated areas, geographically. The case of the vaguely humanoid, but robotic entities, that appear in the Pascagoula, Mississippi abduction, seem to resemble strongly the strange creatures seen hovering about in a case from South Wales in the Pembrokeshire area of Great Britain. Other consistencies are the effects of alleged UFO's on electrical equipment in motor cars, skin rashes and burns on the occupants if exposed to the light source, and time losses and so forth.

Every century, going back to the onset of recorded history, seems to have a few reports of things in the air that startled the observers, and this certainly would be the case as many of the things people report today are natural phenomena, which our technology allows us at least an educated guess at what they might be, but the ancients, naturally would be startled as they had no means of relating them to anything they understood.

Perhaps a small percentage of the sightings in older times were actual aerial craft, given this assumed long observation programme that seems to be underway (or that may soon be drawing to a close).

It is easy to see how ancient astronauts become part of the present phenomena. In particular, the extremely large markings that appear all over the world set out purposely on the ground that, although ancient, attract much of the modern day UFO activity and reports. Not far from the markings and symbols on the Nazca Plain in Peru there is a huge shape, incised in the dry soil of a hillside in the Bay of Pisco (also in Peru), that appears to resemble a three-pronged candlestick, or trident, and has been assigned both of these descriptions. On the ground, it is hard to imagine its size, the width of the centre column being about fifteen feet, two feet deep and fifteen hundred feet long pointing toward the Nazca Plain and those well-known markings.

Even Britain has marking, usually of horses, inscribed on a grand scale on various hills that are, again, seen clearly from the air. The implications are that the ancients all over the world were aware of something in the sky that they seemed anxious to somehow communicate with by use of these huge diagrams and symbols. There does not seem to be any other logical explanation in constructing markings that are indiscernible on the ground.

A certain writer was pilloried for seeing ancient astronauts in everything he wrote about, and to be sure, if one gets too immersed in the topic this can be the case. The same can apply in other matters. The infamous communist 'witch hunts' of the fifties, when a certain US Senator exposed various people for being communists and everyone was seeing 'Reds under the bed'. This same factor applies in UFO 'flaps', which seem to gather momentum after a few consecutive reports, until waves of people are seeing them.

The aforementioned writer travelled extensively and researched his books quite well, but continually referred to 'my theory' when referring to ancient astronauts, when many had preceded him in his particular topic, such as Robert Charroux, Robyn Collins, Raymond Drake, Andrew Tomas and Peter Kolosimo.

There does indeed seem to be enough circumstantial evidence for the theory to stand on its own and countless references, steles, inscriptions, legends and ground markings, all make it clear that the ancients were seeing something that confirmed to either spherical or circular shaped craft, and impressed them enough to record them as best they could.

Many representations of what appear to be suited, helmeted figures appear in ancient cave drawings in the Sahara and in areas of Australia. Strangely, the Saharan drawings seem to indicate pastoral scenes, with abundant cattle and grass to sustain them in areas now dry, dusty and sun baked.

There is no doubt that major climatic changes have occurred in the past, far too profound to have been related to the comparatively small movement occurring during the 'Precession of the Equinoxes', or rotational 'wobble' of the polar axis. When one considers the mile thick ice in the Antarctic, then ancient maps showing every detail and cover which appear now after scientific surveys and 'soundings' to be accurate, many indicators seem to point to a major shift in the Earth's axis and clearly the Antarctic coast was once ice-free. If the Antarctic ice does ever melt away it may be surprising what we might discover there – perhaps evidence of ancient civilisations.

The detail of the Antarctic mapping also implies the necessity to utilise the maps for journeys to and from the region, otherwise, unless they were the result of some vast world geographic survey (which seems unlikely), there would have been little point in the ancient seafarers taking so much trouble to produce them (assuming the mapping was done from the sea and not the 'air' which the well known 'Pirie Reis' map is said to imply). Were there settlements in the region at the time when weather conditions were not so severe, making it advantageous to have good maps for trading arrivals and departures? Deep core drilling in the region have indicated this.

The Earth literally teems with unanswered questions on constructions, artefacts, drawings, ground inscriptions, maps, enigmatic stone constructions and so forth, and it may be only necessary to find a key in the understanding of some of it for the rest to fall easily into place, but we are seemingly constrained by certain preconceived notions with regard to our written history and the chronological order we have assigned to it all.

Many symmetrical constructions, which appear to be buildings that once stood on dry land exist off the east coast of America. Sometimes they are spotted from aircraft in calm conditions and are difficult to find again. Whether they are related to the mythical lost Atlantis or are just settlements that were submerged during the melting of the last Ice Age, are not known. If a large mass did sink then the Gulf Stream would flow north to melt the ice and raise the sea levels. The undersea 'finds' do not seem to be natural formations and in any event, divers have made some extremely interesting discoveries in the sea areas commonly known as being part of the so-called 'Bermuda Triangle', and among them have been caves with evidence of stalagmites and stalactites inside them, clearly indicating that they were once above the surface of the current sea level. With regard to items such as the symmetrical artefacts, they seem to indicate buildings and settlements, columns, blocks and Graeco-Roman type constructions. Some of them appear to resemble those existing and being reclaimed from the ever-encroaching jungle-type fauna of South America, where there may be many more hidden wonders awaiting discovery. When one takes a long look at all the fine ancient constructions that we often marvel at with awe, it would appear that our knowledge regarding the ability to construct buildings to really last through time, has long been lost to us, and various places on Earth in the dim past must have been a sight to behold, and no doubt inspired the Greeks and Romans, and it has been suggested that 'cosmic guidance' with superior knowledge in regard to building in accordance with 'Earth's forces' was imparted to the ancients by 'initiators' or 'special' beings.

However. I believe we do in some cases underestimate the skills and knowledge of the ancients and their capabilities, but it is a fact that many Earthly legends speak of these strange 'teachers' particularly the South American variety where these special beings left after imparting all this wisdom, promising to one day return. The fact that many UFO sightings have been seen in areas of great antiquity may simply mean that if these aerial craft are manned by extra terrestrial entities, they may not be in any way connected with them and could simply mean that 'they' are just as interested as we are in them.

With regard to the really ancient monuments from the Neolithic period, and this would include the Menhirs, Dolmens and, perhaps more purpose built constructions like Stonehenge, those that have studied them intently seem to suggest that they are not haphazardly situated and the builders understood the secrets of 'Earth forces' and so called 'Ley' lines, and perhaps there is some connection with the ancient art of divining and locating underground water courses, some of which are said to flow like mighty surface rivers. It is well known that liquid flowing through pipes causes enormous static electric build-up, which is why aircraft refuelling hoses have to be 'earthed' by a plug attached by a cable to the hose and inserted into the aircraft to dissipate the charge through the aircraft to the ground to prevent sparking. Whenever you turn on your tap static electricity is immediately built up, plumbers are aware of it and 'earth' the copper pipes by the use of a wire earthing connection one can see under the cupboard ware of the kitchen sink.

Any fast flowing underground watercourse would, no doubt, be generating static electricity into the surrounding earth at quite a magnitude. These remaining standing stones seem not only to be 'linked', in a purposeful network in Britain, but also into Europe as well. Our general knowledge of them is so sketchy that any hypothetical aliens would not need to know much to be ahead of us in regard to their knowledge of them.

With regard to older encounters, seemingly of the 'third kind', it would appear that certain manuscripts exist in the British Museum regarding various conversations between John Dee, a scholar, astrologist and one-time intelligence agent to Queen Elizabeth the First, and 'beings of another sphere'.

It seems that in 1581 Dr Dee was contacted during his studies by an entity in the middle of a dazzling light. The being introduced itself as 'Angel Uriel'. At his alarm and amazement the being merely smiled and handed Dr Dee an artefact in the form of a convex crystal and informed him that he could use it to converse with beings from 'another sphere'. It would appear that John Dee involved a certain Edward Kelly to witness it all, and between them they managed to learn a complete language known as 'Enochian' and that this language was used by later occultists as a powerful tool of their trade, so to speak, such as Alaister Crowley.

Another interesting footnote to this is that John Dee, during his post as the intelligence agent to Queen Elizabeth, always signed his reports from abroad with the number 007. Clearly, it seems Ian Fleming was familiar with the history and writing of these characters from the Renaissance.

A poet of the Italian Renaissance by the name of Ludovico Ariosto, who died shortly after John Dee was born, seemed to be writing of aerial phenomena when he wrote in one of his stanzas, 'Proud demons, sailing the heavens in great ships of glass', and in another, 'up to the starry sphere with swift ascent, the wizard soars'. Part of another stanza read, perhaps it was a bird, but when or where another bird resembling this was seen, I know not ....'. What can we deduce from another that reads, 'The chariot, towering, threads the fiery sphere and rises thence into the lunar reign. This in its larger part, they find as clear as polished steel when undefiled by stain'.

Other areas of the arts from this particular period, seemed equally as preoccupied in celestial depictions. A Venetian painting from 1538 of Christ on Mount Tabor, shows him clearly sitting inside an egg-shaped craft with flames at the base and attended by angels whose haloes look exactly like space helmets, or breathing domes.

Some years later, perhaps even William Shakespeare himself saw something he could not explain, which is now forever logged in his play *King John* of 1596, when the character Hubert said, "My Lord, they say five moons were seen tonight; four fixed and the other did whirl about the other four in wondrous motion". It seems that this period was quite a busy time with regard to aerial phenomena.

While the above extracts are the obvious ones, no doubt many writings throughout history could have sentences highlighted that seem to be relating to aerial phenomena, but there then exists the aforesaid danger of seeing them practically wherever we choose in any historical writings.

With further regard to the ancient sighting, we could mention an event of 698 BC, when the Prophet Isaiah stated, 'Who are these that fly as a cloud and as the doves to their windows' (UFO's returning to a mother ship?). Other strange anomalies and reports from the past consist of an event in 1790 France, of a close encounter with a craft that was seen to be consumed, or 'self-destruct', after landing on Earth. Shortly before it burned up, a being, in close fitting attire, emerged from it and ran into the woods.

A Police Inspector was despatched from Paris to investigate the case at Alençon, where some peasants had been alarmed at a 'great globe' that alighted and set fire to some grass. People came running from all directions to see it and among them were a physician, the mayor and other local dignitaries. This event would be really quite easy to explain as simply 'ball lighting' except for the fact that it had an occupant.

The being that emerged from it uttered some incomprehensible language before running into the woods (perhaps it was 'keep back' in alien parlance), then the sphere silently exploded and the pieces all consumed themselves and turned to powder which dispersed in the wind, perhaps it was formed of antimatter that had become unstable causing the occupant to make an emergency landing.

Perhaps the bones of at least one alien entity now reside in the Earth, as the being was never found. On the other hand, he may have set up some kind of rescue beacon signalling device like that which was so amusingly portrayed in ET (The Movie), to attract a rescue craft, to pick him up. It does not seem likely that anyone would conduct such a strange story.

Quite naturally, this report when sent to the Paris Academy of Sciences, was not believed. Even with our comparative sophistication in celestial occurrences, we ourselves would find it hard to believe. Nevertheless, many strange accounts and occurrences exist in the historical records of all nations.

Have we an explanation yet for the missing soldiers alleged to have disappeared into a 'cloud' in the Dardanelles campaign in Turkey in August 1915? This event is supposed to have swallowed up a British regiment of a few hundred men. The 'cloud' lifted off and joined others heading east. It was estimated to have been around 240 meters long and 60 meters thick and of a strange greyish colour, but with a dense appearance. Turkey expressed no knowledge of the missing soldiers when their return was demanded in 1918 by the British War Office, who naturally assumed they were POW's. The M.O.D. would not like to admit to this event due to its high 'strangeness' and could easily deposit badges insignia, uniform and even bones there if they so wished.

Perhaps one of the strangest of all discoveries was the mummified body of a tiny man found in a cave in Caspar, Wyoming in 1932, and there was a report of an X-ray of the creature in the National Enquirer of February 1968. The X-ray was allegedly carried out by the Anthropological Department of Harvard University. The creature was said to have been around fourteen inches tall and weighed around twelve ounces. It had a full set of teeth, had all the tiny bones in place and was said to be humanoid and around sixty-five years old. This sound more like an April fool joke than fact, but many strange stories exist in various countries' records and one assumes some of them must be based on fact, and as this case in particular, seems well documented, it can be checked out.

The strangest anomalies are those occurrences, such as cattle mutilations, that exist in old reports as well as still occurring today. Certain veterinary surgeons say birds are responsible with their precise incisions of razor sharp beaks in surgically eating away the flesh, but the farmers that have witnessed the precisely incised and removed portions evidently do not seem to agree with this thesis at the present time, and many events are seemingly associated with the ongoing UFO phenomena with craft being observed before certain events occur.

A fantastic report from 19th century America had a certain Alexander Hamilton, a Kansas farmer and member of the House of Representatives, together with his tenant G D Heslip, witnessing "A great cigar shaped craft, possibly 300 feet long, and a lower part made of glass or transparent material, occupied by six strong beings", that had lifted up one of his cows bodily on a cable of red material and moved off with it hanging below. Later the hide, legs and head were found in another field and no traces were found in the soft ground. Cattle mutilations, sometimes surgically precise, still occur today. We should mention that balloons existed at that time and this could simply be an (air) rustling case for at least two months meat supply for the occupants.

The long record of such observations does not imply beings wishing to subdue the Earth, but any earth-like worlds that may now be analysing our communication emissions, may find them a godsend. They may have been searching fruitlessly for a new home, perhaps having caused irreparable ecological damage to their own world or having seriously depleted its resources. They may be tempted to, (after looking us over) take the Earth for themselves. In addition, this, as said, is a more serious aspect of announcing our presence so freely to the universe.

Of course, they may prefer to achieve their aims by other means than annihilating us. In any case, they could not evacuate their entire population, but lists may well be in the process of being drawn up to ensure at least a chance of their lifeform to survive elsewhere, and a probe could be here already or on its way ahead of the coming 'Armada'. This will cut down their observation and analysis time of the intelligence source and allow them the time to formulate plans on how to deal with us. In any event, it may be a one-way journey for them. They would have nothing to lose and everything to gain. Such entities far from being belligerent and covering the Earth by force, could arrive totally defenceless, simply landing on unknown and perhaps hostile shores, putting themselves at the mercy of the natives.

Well we can always hope can't we? Of course, they could arrive like the Viking hordes.

When the science of radio astronomy began sky searching in earnest, and organised S.E.T.I. programmes were initiated, there were a number of false alarms, until certain phenomena, now commonly known about, were determined as being responsible for what appeared to be purposeful signals. Among these were pulsars, close orbital binaries, blocking each other's radiations intermittently, the neutron stars and so forth. In other words, the wheat had to be separated from the chaff so to speak, then the teams could settle down to look for real evidence of purposeful intelligent transmissions.

As long ago as 1960 when Dr Frank Drake of the US National Astronomy Observatory, began Project Ozma, almost immediately after switching on his equipment and eventually lining up on a sun-like star, Epsilon Eridani, 10.8 light years away, he seemed to be receiving a coherent signal and naturally, the project team became quite excited. Sure enough, it turned out to be a false alarm and was traced to a powerful emission from an Earth borne source. This was not to be the only false alarm and one imagines nowadays, they become a little more wary, and it would be something quite profound that would have to occur, before it raised the odd eyebrow. Of course, it is entirely possible that intelligent radiations were streaming into earth space before we had the technological capability to receive them.

The false alarms have all been well documented and the search continues. It was deduced, quite naturally, that any intelligence would transmit on a common frequency, such as that of hydrogen, and these factors have all been considered and many, many emission lines have been scanned. As most people know the spectrum or light from a star or colour band the black lines in the colour band represent the stars composition, hydrogen helium etc., and each has its own wavelength or emission frequency.

Radio astronomers from the University of California detected radio signals from space, which had patterns totally different from anything in the radio spectrum, and they fluctuated in time scales of hours or less and were christened 'mysterium'.

Dr Robert Jastrow, one time director of The Goddard Institute of Space Sciences at NASA believed that beings in other solar systems have been analysing our radio emissions for years.

Dr M.A. Mercer who was a physics lecturer at Southampton University, said "There is the case of long delayed radio transmission 'echoes' investigated about thirty years ago, but never explained. It is entertaining to speculate that they may have been picked up by an (alien) exploring probe and relayed back to us". They must have bounced back by hitting something, or perhaps received and re-transmitted.

There has even been speculation that some super advanced intelligence could have artificially 'created' the pulsars as celestial navigation beacons, simply because their signals are so regular and precise. For example one every 1.3372795 seconds and an accuracy of 1 point in 100,00,000 and this equals that of atomic clocks. One X-ray source pulsar beams every 1.24 seconds for 9 days, shuts off for 27 days, then recommences the same cycle again. Dr Frank Drake, who could be described as a pioneer of radio astronomy sky search attempts to contact extra terrestrial intelligences, said "There is no known reason for a 'star' to behave in this strange manner". Professor Alan Barret, who was a member of the Massachusetts Institute of Technology, wondered if we had stumbled on a vast interstellar communications network.

What of the enigmatic face on Mars? It is reported to be 'bilaterally symmetrical'. However, NASA is said to have stated that a further set of pictures were taken and that no 'face' appears, yet they have not released the pictures. However, they have now successfully 'debunked' it by releasing the structure shown as a natural feature but to others are suspicious of what they see as a contrived image.

Since they were happy enough for the original shots to appear worldwide and in numerous magazines and books, why be reticent to release the latest pictures? Of course, we know they would not resort to subterfuge and release pictures of the area with the edifice purposely obliterated, because (if for no other reason) so many people within the Organisation would know if that occurred and most likely, there would be a leak. Therefore, many might conclude that they have confirmed something quite profound. It has to be admitted that many strange things encountered by the rover and from ariel craft have been shown on such TV programmes as the NASA mystery files.

It seems that there are other intriguing things on the Marian surface to explain. As well as the 'face-like' edifice, reference has also been made to a system of 'pyramidal' structures appearing on some of the Mars photographs and they are said to have been computed as being about half a mile in height and are described as being five-sided, with clearly defined edges, suggesting, by their appearance, of being objects of special design and construction.

Important people involved in NASA projects are reputed to have said, "They warrant a closer look", and that they seem to have some 'engineered design'. It seems to be the case that until our manned landers arrive on Mars and the astronauts walk up and touch them, the mysterious edifices will just have to join all the other mysterious and unexplained anomalies that are voluminous enough to be the subject of complete books on these topics that fill our library shelves in the paranormal section. The pinnacle of the Egyptian pyramid Khufu, emits an electronic force and pyramid structures exist worldwide, stones used in other constructions such as the obelisks of spires contain Chrystal and excellent radio receiving element.

It would appear that many UFO sightings have occurred at the 'junctions' or crossing points of the aforementioned leylines. It seems that certain dowsers have detected these 'leys' or energy lines and, since dowsers have been involved, perhaps they are rushing underground water courses, causing some form of energy manifestation of a static electric, or electromagnetic nature that are curious and quite detectable to alleged UFO entities who may have mastered complete control of such power sources for their craft and may be able to display the lines like a 'grid' system on their instruments. Major and most enigmatic stone structures when studied geographically show a 'connect' pattern along the latitudinal meridian lines all over the world.

There are energy fields all around us, even the human body itself has an 'aura' – another word for an energy field – said to be detectable and seen by certain gifted individuals and alleged to have been shown in 'Kirlian' photography, which may simply be 'mass field radiation', that everything with mass is said to possess, and it has been proposed that the UFO power sources with their ability to disappear when threatened by intercepting aircraft, are able to impose an 'anti-mass field generation' on their craft by the revolving rim of the craft being connected to banks of anti-mass field generators within the core of the craft, thereby producing this ability to become massless, and therefore invisible, and this ability would also allow them to make abrupt manoeuvres and right angled turns, being unaffected by gravity and inertia.

Did the ancients know of this bodily energy field in the ancient art of acupuncture? The experiments in the aforesaid Kirlian photography seem to suggest that if scar tissue, or perhaps a small cut or wound was apparent on the

Photographed hand, or a withered area on a leaf, for example (which have also displayed this 'aura'), the affected area had an obviously diminished brightness in that particular part of the aura.

With regard to the very ancient art of acupuncture, and, as manifested in the case of inserting needles into affected areas in people subjecting themselves for treatment in the process of this so called 'fringe' technique, they could be having their 'aura' fully restored and subsequently their general health. When I mentioned the energy fields that are all around us, anyone who has stood under the crackling buzz of energy apparent in power lines will be aware of this 'field', and if it could be seen, how far reaching it was there may be some alarm. It could have been having quite profound effects on human tissue going unnoticed until our fairly recent awareness of the dangers of it.

Certain correlations with cancerous brain tumours have been made with people who have developed them being subsequently found to have been prolonged to excessive exposure to the fields by having power lines passing directly over their garden and science is aware of this correlation. Anything with mass probably has an energy field as mentioned in the aforesaid possible UFO power sources neutralising mass by anti-mass generation, freeing their craft from inertial and gravitational constraints, as mentioned on the previous page.

There are probably as many underground rivers (perhaps more) than there are on the surface, and if the alleged UFO phenomena can detect energy output manifesting itself, they may see on their supposed detection apparatus as said, the complete grid system of these energy, or 'Leylines'. How could the ancients know of them? It would appear that many of the 'Leys' go straight from one ancient monument or church to another in a purposeful straight line and, of course, many old churches were built on much older constructions, sometimes intentionally to suppress the ancient cults and beliefs during the Christianisation of Britain.

Similarly, the feasts and special events in the ancient calendar and seasons had Christian events gradually super-imposed over them to ease the trauma of change for the ancients. One again, the evidence seems to point to this knowledge purposely destroyed or suppressed, which seems to be a recurring factor throughout the world history as a whole.

It is hard to state or decide whether the human brain's extra activity regions, that manifest themselves in a small portion of the population, are a feature of the brain further evolving (and eventually everyone will have so-called paranormal capabilities), or whether they are lost powers suppressed and unused, when sheer survival was uppermost in the mind of the ancients.

Medical science has long been aware of the possible link between 'brainpower', and that a positive outlook with regard to one's health makes for fitter people, than heading for the doctor's surgery after the first sneeze and worrying over every bodily ache. 'Stigmata' is beginning to be suspected as a mental condition rather than miraculous by the Church (and possibly the cures at Lourdes). A Russian woman has demonstrated brainpower by moving ball bearings about on a plate by mental thought power. Then there is a strange case of 'thoughtography'. Certain gifted individuals can stare into a camera lens and produce what they are thinking of on a photographic plate. Studies of these occurrences are being undertaken by many countries with the same old motivation, how it could be utilised for military purpose.

Poltergeist activity is little understood, but whereas in olden times it was clearly assigned as 'noisy spirits', as the name suggests to those with any understanding of the German language, today it is also being considered as a mental manifestation by one of the 'victims' and likely to be caused by that person. That is the general belief but such activity has occurred when the occupant is not in the property or as in a recent case when a paranormal group (Most Haunted) found a chair perched precariously on top of a door.

Clandestine security organisations east and west have long recognised the enormous potential of using or developing the powers of thought transference that seem to have no distance barriers (with regard to Earth distance, at any rate), as many cases are on record of a person in a moment of tragedy, far from home, appearing to a relative at the moment of death, the most famous being that of Admiral Sir George Tyron, appearing to his wife, Lady Tyron, by walking through the drawing room as he drowned on June 22nd 1893 in the Mediterranean. With his final strong last thoughts of wife and home, he appears to have projected his image there. All of these advantageous powers may have been mastered by E.T.'s long ago.

Telepathy also seems either a latent power or showing signs of developing in the brain. Again, with regard to this power, Earthly distances (and perhaps even cosmic distances, when fully developed) seem to be no barrier. The power seems more pronounced in twins and physical experiences, such as pain or illness, seem somehow to be detected by the member of the set of twins, even in some cases, from halfway around the world. All these features may have been part of the initial insertion of alien genetic material that was installed in the human many millennia ago, according to ancient astronaut theory.

So then, if such things as the aforementioned and other powers, such as dowsing, levitation, theory psychic healing and mind over matter, the ability to project one's image in out-of-the-body experiences, etc., are now being experienced and manifested in humans that are alleged to have so quickly reached our currently evolved state in a few million years, what would humans be capable if it we had been around for so long as the time assigned to the simian lineage, i.e. the earliest 'Pongid apes'? Given the cosmically short time span of another thousand years, humans may have mastered all such powers. What would beings having, perhaps a million years of development more than use, be capable of? Are we witnessing some of their abilities in our skies today?

As said earlier in the work, UFO activity is not simply confined to one aspect of earthly activity, such as our defence establishment and capabilities. If it was so confined, we could read sinister connotations into such activity.

However, 'they' seem to be interested in every aspect of Earth and its creatures, and cows seem to be high on that list of interests, but as far as I know, no human beings have been found with neatly incised portions of the body form missing. Although having said that, many police files could be produced to show some very bizarre cases of people being found drained of blood, and other strange occurrences, which can usually be found to be caused by human hands, and many people with an admitted proclivity for drinking blood do, in fact, exist. Hundreds of them are said to live in America and there are allegedly thirty-six registered drinkers of human blood in the area of Los Angles in the USA alone. No doubt, many exist in other countries who do not wish others to know of their strange fetish.

It is probably not a new phenomenon as the actual character from the Middle Ages, 'Dracula' or Vladtepes, a bloodthirsty General from Transylvania, inspired Bram Stoker in his novel *Dracula*, allegedly pursued a gruesome interest in blood. It would appear that there is a very wide spectrum of human brain activity in either the positive or negative mode, and if we can see the simple necessity to locate the possible neurones responsible for the dark side and by 'adjusting' them, cure these mental afflictions instead of locking people away and simply administering calming drugs that only suppress the problem rather than cure it, then we may one day have the medical scientific ability to enhance our own mental activity, just as, maybe, the goal of the alleged beings, said to be observing us, and it may just be true that 'they' could do it for us, should they ever be able to win over the hearts and minds of the world populations in order to submit to such treatment, which would, in modern parlance, be a 'big ask', given our proclivity for aggressive reaction and violent response. These features in the human are the greatest obstacle to civilised advancement.

It would appear then, that if we accept the reality of the phenomena in our skies (and more and more convincing cases come to light almost every day) then we must allow for a final plan that 'they' must have in mind. Surely, they would not intend to dart about our skies indefinitely. There is no doubt that to form an acceptable plan is their greatest dilemma.

Recently an aircraft on landing approach to Manchester, England, had the crew observing a wedge shaped craft streaking by their aircraft, that was described as having a distinctly unearthly design and was near enough to make the co-pilot instinctively duck, and the mandatory 'Air miss' was filed. It is said that few pilots these days report them (for their own logical reasons), but this, it would appear, was a very close encounter and incidentally had one or two ground observers tending to corroborate the report. It may be that these close encounters, seemingly getting ever closer, are part of their preparation plan.

Most certainly, among the retained files that obviously exist in the Registries of the various security agencies, would be the famous Rendlesham Forest encounter and would be filed away in British and American defence organisations, since, though it happened on, or very close to a US air base, that air base was in the UK. It occurred in December 1980. It involved USAF and RAF personnel. The usual calming statements were put out that the encounter was not seen as a threat to security of our air space (or air base).

Even the commanding office of the base, with the rank of colonel, was actively involved in the encounter, so the authorities had to listen up a little more than they would if the event was reported by a couple of enlisted men. The MOD in London just as, no doubt, the Pentagon, is in full possession of the facts.

The Rendlesham Forest might be seen as an ideal location for clandestine spying activity by ETs (or earthly personnel). 'They' may be somewhat interested in military activity, as two air bases exist at either end of three miles of forest. They encountered a triangular shaped object, brilliant red and pulsating, and the phenomena disappeared and returned a couple of days later.

The events were witnessed by people competent in identifying all known earthly flying craft. On this last point, of the shape of the object, was said to be like a diamond, but whatever shape is mentioned, that form would be on a file somewhere. Many people believe that some of the objects are from US black projects, resulting from crash/retrieved ET craft now being studied in a remote region of the USA. The authorities would prefer that people believe they are UFO's which they can deride or debunk, rather than admit to their secret projects.

If we do assume a long and patient 'alien' observation period of Earth, we have to consider the obvious fact that 'they' may not simply 'go away', 'earth's' peoples must be made aware of the ultimate purpose of it all one day. Human origins are still largely a mystery and the theory that mystical beings may have been involved in the evolution process of our species cannot be ignored.

The theory that an advanced ET race came to Earth and created men in their image, rather like a divine creator is unpalatable and disturbing to devout Christians, but there is one thing they must confront and that is their natural logical thought process that they cannot simply switch off. They cannot just unquestionably accept creation as per Genesis, but many do so, but they have the right to ask where is the 'definitive proof 'that humans were once apes'? Good question, only theory exists, in place of evidence. They may say 'science has proved through the studies of mitochondrial DNA that is only inherited through the female, it must therefore have an original 'earth mother' who Christians would say was 'Eve'? However, what were her racial characteristics? They could ask where is the proof of ET creating humans, again only circumstantial evidence exists. The UFO phenomena is undeniable, the studies of the human genome offer many clues with rare and strange genes discovered that compel us to try to identify their origins and who the possible 'donors' where. The question regarding the enormous intellectual capacity in certain gifted humans cannot be answered by our present understanding of evolutionary and natural selection processes, but it could be answered by belief in an almighty creative God, but then we must believe in the mind stretching Genesis story. The belief in human evolution is not compelling as poof but the vital bone fossil evidence could be discovered at any time or never. On the other hand, a religious 'second coming' must do so to vindicate the widespread belief in it. Indeed, so must the ET entities (where the evidence does suggest their lengthy presence on Earth) appear in some manner to prove their past deeds in order to answer the 'human question' regarding our puzzling origins? Not only would this be a disturbing and serious cultural shock, but it would cause an enormous discharge of beliefs, history, written works, documents and accepted theories 'wonderful' many would say, no more wondering, false beliefs and questions, at last the only true 'REVELATIONS' are portrayed.

It is rather obvious that since the various authorities go to such lengths to assure the public there is nothing to the UFO phenomena, they certainly would not wish it to be known they were authorising groups of supposedly qualified people to investigate them, but since the phenomena is so, prevalent, worldwide and disturbing to many, it cannot possibly be ignored, therefore it must be under intense analysis.

It is doubtful that the hardworking and sincere UFO voluntary investigation groups are made up of any impressively qualified personnel, but rather the opposite.   They are probably everyday working people of the various persuasions, but with plenty of common sense, which would clearly be the main prerequisite to separate the obvious hoaxes and more silly events from the seemingly genuine ones. Having said that, they do have ex-military and a few PhD's among their ranks, the former having had UFO's experiences themselves.

At first, it seems strange that astronomers should be the people who seem mostly to distance themselves from UFO activity when most people would assume it to be the other way around.   However, when one considers that apart from the amateur group mostly lining up on the Moon and the planets, the professionals are virtually queuing up for viewing time and look at one particular aspect at a time in their science, that is other galaxies, radio sources, pulsars, gas clouds and so forth, and are largely pre-occupied with solving the problems of the dark matter and so on.   They are looking at one specific target and anything fairly close by, passing their field of view, would go virtually unnoticed. It is the people that look up at the sky generally for lengthy periods, and there are not many of them if we think about it that would be most likely to see something and those whose attention is diverted by something distracting them in their general field of vision and who would be best positioned to be able to see these manifestations, and the obvious people who would fall into these categories would be patrolling security guards, policemen, pilots by night and day, even Gypsies and New Age Travellers, holidaymakers in the mountains and areas of natures beauty, particularly in the USA.

In other work, I mentioned my encounter (in a middle eastern hotspot, when at a remote guard location on an airfield) with a large metallic looking circular object silently passing overhead that caught any attention and caused me to look up. I was familiarising myself with the base and its security set up.

After gauging it to be about three times the size of the full Moon and watching fascinated as it went quickly out of sight a few seconds later I heard a hissing and clattering noise from a nearby building which I later found was the Meteorological Office, and I knew that it was a weather balloon, the reason offered frequently (perhaps too frequently) to explain UFO's but it taught me that it is a valid one.

Many ufologists that deride this rather constantly trotted out reason for UFO phenomena probably have never seen one, but at night, I can assure them that they are extremely 'UFO-like'. They can also move quite fast when gaining altitude and picking up high altitude air streams. However, I have no explanation of the strikingly blue disc I observed that traversed the sky in seconds, not dimming or brightening, it was about the size of a 10 pence piece held at arm's length, streaking north.

With regard to the abduction phenomena, seemingly on the increase and in relation to the energy fields and the human aura previously mentioned, an interesting case occurred near Stockholm in 1974, when a person walking along a dark road was surrounded by an intense beam of light and was apparently drawn up into an overhead craft. His experience, when later drawn out by hypnotic regression methods, conformed to the usual events alleged to have taken place on other classical abduction scenarios, i.e. the usual medical examination and so forth, the suggestion of telepathic communication and so forth, significantly he does not seem to have been telepathically urged to forget it, as in earlier abduction cases.

He had certain wounds, the positions of which coincided with certain statements of inserted needles or probes, but rather amazingly he seemed to have become somehow 'magnetised' and could affect compass needles and was tested with his ability to affect iron filings, much like using a magnet to produce the 'field' effects in school experiments.

Of course, we all have some manifestations in this regard with some people not being able to wear wristwatches or sparking effects when shaking hands and so forth. I think the strangest thing I have witnessed was a person I knew of that could not put a ballpoint pen in his pocket without the ink erupting all over his pocket. However, whether this strange magnetic effect was apparent in all of the alleged abductees (or was looked for) is not known. It could be significant if all alleged abductees where tested for this effect, but it may not be obvious after a certain length of time.

It has to be said that 90-95% of all UFO reports are classified as 'explained' but when one considers the sheer volume of the reports from worldwide sources, the remaining amount of 5% of unsolved reports amount to a lot of cases, for which we do not have the ready answers. In any case too many events are quickly classified as solved on the flimsiest of evidence, project 'Blue Book' was renowned for this.

The same percentage could be used in an analogy of the universe, being largely made up in its composition of hydrogen but there still remains a lot of hard material out there from rocks, comets, dark matter burned out suns and maybe even free worlds, material ranging across a wide spectrum from dust particles to black holes. Earlier where I mentioned certain aircraft as being possibly influenced in design by captured, crashed UFO craft, retrieved for study of power sources and design, it is interesting to note that a certain Lt. Colonel George Edwards, one time serving officer in the USAF, stated that the US Air Force was secretly testing and flying captured alien space craft in remote locations. It is hardly likely he would risk his reputation for being prosecuted for breaking the Official Secrets Act. This could be black propaganda to divert attention from the fear of UFO's.

However, many UFO reports have been made when the aircraft, such as the Stealth Bomber was being flight-tested and would remain unexplained at the time, because no one was allowed, or was even willing to explain them, for obvious reasons. If the military had the ability to capture UFO, craft the public would be aware they are vulnerable and be therefore less afraid. However, encounters by aircraft that are purposely directed towards unidentified aerial craft, that are the result of blips appearing on radar screens and subsequently identified by the aircraft sent in to investigate and appear as silvery, disc-shaped, and travelling at 1,000 miles per hour can be, it would seem, little else than a bona fide UFO, but as said, we always have to keep in mind that unidentified does not necessarily mean 'extra terrestrial'.

Just such a case as described, occurred in 1951 with a T33 Jet from an Air Force base in Delaware USA, with two crew members and ground radar simultaneously confirming the encounter. Certain reasons for this sighting among those usually trotted out can be immediately eliminated, such as space hardware re-entering, as we, had not 'entered' any in 1951, (apart from the rapid up and down flights of V2 rockets).

As mentioned, when senior military men break the Official Secrets Act it seems the case for black propaganda is quite strong. Senior military people would be aware the population would be more comfortable when hearing about US advanced capabilities than worrying about UFO's. Such officers have clearly been encouraged to make such statements.

If a thousand miles per hour seems fast for 1951, in the following year 1952, seven distinct radar returns or blips in formation were picked up by ground radar moving at 7,000 miles per hour. In these cases, the equipment is always checked and in this particular case, the equipment proved serviceable. The above happened at Washington National Airport, Air Traffic Control Centre and showed no regard for restricted airspace when virtually passing over the Whitehouse, assuming ET's were aware of such restrictions. It may have been a purposeful to act to "let the leader of the most powerful nation on Earth see us".

With regard to evidence of landings of alleged UFO craft, an Australian farmer reported a huge depression in one of his fields, a subsequent calculation seemed to indicate an object of some 600 tons, and a width of 500 feet had settled there. However, the precise crop circles reported worldwide with evermore-intricate patterns, seem to have no inner depression and whereas some are hoaxes, they certainly are not 'landing impressions', the greatest mystery is why the perpetrators are never caught in the act (if they are human).

Not to appear too cynical with regard to their formation, they mostly appear in the warm summer evenings, when the students are on vacation and, as everyone knows, the books on them are numerous in the local library – a popular haunt of the students. In any case, people have actually admitted to producing them and have demonstrated their methods of carrying out the work. However, the above question remains. Farmers spend more time awake than asleep and some patterns are intricate and extensive and would take a lengthy period of time to produce. To return to the UFO phenomena a certain RAF Wing Commander must have felt quite insulted when on his encounter with a UFO he was guided toward by ground radar was explained away by the authorities as the planet Jupiter. The ground personnel had been alerted to it by other aircraft and had a radar return of it. The object manoeuvred rapidly to avoid thorough identification and was reported to have red, white and green lights. Of course, these are the colours of normal earthly identification lights on our aircraft. If it was an earthly craft, a check of the flight plan submissions and on known air traffic in the area would have identified it readily enough. Therefore, it either belonged to a foreign power or was a genuine UFO. These kind of cases and the treatment of those involved in the sighting and chase, make it clear that many more cases go unreported, simply to avoid humiliation and disbelief.

But how the planet Jupiter could give a radar return, or dart about, was an insult to the pilot's intelligence, who made the encounter, as said, it is this kind of 'casual' debunking that prevents many pilots from reporting things they observe (and suit the people well, who have to do the debunking), the less cases the better, the less half hearted investigations and findings suitable explanations the better, although some explanations appear quite unsuitable.

Regarding the 'debunking' and cover ups, (that the ufologists insist are still happening), it is interesting to note that the CIA, who vehemently denied any involvement in UFO's or their investigation initially were eventually, due to the Freedom of Information Act, made to cough up 40,000 pages of documents. Many of these, of course, would have the usual purposely made, major felt tip deletions, which as said, would be quite undeleted in the retained copies. This is an amazing amount of data to be withheld from the public and of course, highlights, the fact they were lying when saying they did not have them.

Can we assume then, that there may be a possible conclusion to the alleged 'UFOs observing us' phenomena in the near future? Not only because of the many predictions made for dramatic events to occur at the end of the last millennium, but it was a nice round date to choose for any entities that may have been responsible for another event of two thousand years ago in the "was Jesus an astronaut?" (also unanswered) question. A couple of millennia of E.T. presence implies more of a responsibly than a threat by alleged E.T.'s. when we postulate that Earth may have been under observation since biblical times, we may be underestimating the observation time. If cave paintings show discs and orbs and figures in what appear to be space suits from thousands of years ago, we must conclude when noticing how accurately they have portrayed run of the mill objects and life forms, such as cattle, that they have accurately recorded the other things they have observed, that must have impressed them enough to ensure their depiction of them, on the basis that 'a picture is worth a thousand words'.

If we ourselves can formulate the opinion that in spite of all our preconditioning (it has actually been suggested that producing the film *Close Encounters* was a result of a 'preparation suggestion'), we are most assuredly not mentally ready for traumatic revelations or encounters with extraterrestrial entities. Then given their assumed advanced intellect, they should also conclude that we are not ready, so perhaps we have another millennia of encounters to endure before such revelations. Coming back to the time dilation theory that length of time would not be as long for them as they busied themselves elsewhere in the cosmos.

A logical process of deduction seems to indicate that such entities would not be infinitely benign and patient and would wish eventually for a conclusion to other operations with regard to planet Earth.

We seem to be extremely interested in the planet Mars and definite plans are afoot to go there but there are enormous problems to overcome. Mars has insufficient magnetic fields to divert the 'solar wind' and its deadly radiation and any dwellings manufactures would, if one the surface, surely be subject to it and of course meteorites.

'They' may have some startling revelations about the planet Mars to impart to us before we commence any despoiling operations there, which will certainly occur (given our freedom to pursue) before the end of the millennia. In addition, with the aforesaid dangers mentioned, underground dwellings may be a better option, but that would entail transporting heavy plant equipment there.

One thing is for sure and that is, that reports of UFOs will not diminish or go away either they are explained or the motivating body reveals itself to us.

when our military aircraft crash, unless the reason is obvious such as engine failure or collision, we are told the matter is 'under investigation'. With regard to the Air Force, certain 'restricted' magazines appear that explain the technical faults that may have caused them. The civilian counterpart would be something like 'Flight' magazine, but many an aircraft has been lost preventing any subsequent enquiry regarding the cause, in the Bermuda Triangle for example, for which E.T. has had to bear the blame.

It is known that aircraft have been lost in various encounters with aerial phenomena but this information would not be freely given out. the most famous was probably the Captain Mantell pursuit of an 'enormous disc', which was also described as metallic looking. The result of his death was said to be due to oxygen starvation at great height, through chasing 'the Planet Venus'. As the pilot is dead, he cannot refute the insult. At best, a person with the keenest eyes would not be able to ascertain a 'disc' shape to Venus. It would be seen as a very bright star and not tremendous, metallic looking etc., and an experienced pilot would certainly be aware of the planet Venus from navigational training missions.

Furthermore, though not going into details on the subject, a certain commanding general of the Air Defence Command in the USAF stated, "We have to take them seriously when considering how many men and machines we have lost in pursing them". His name is (or was) General Benjamin Chidlaw. This is a far cry from black propaganda or any purposeful debunking. One wonders what this chief of staff thought about his revelations, as they mostly stringently deny their presence.

The aforementioned Air Force Captain would also be quite familiar with the other reason given for his demise, that he was chasing a weather balloon, but when one sees one in flight for the first time they are decidedly 'UFO looking', but any of the bases most likely to have released such a balloon, as would appear in that specific area at that particular time, have not confirmed such a release and records of such operations as balloon releases are compiled and retained, in any case modern jets could pursue, catch up, circle and photograph any weather balloon in no time at all. With modern meteorological equipment, the need for weather balloons has diminished if not terminated.

The concept of an alien intelligence having sent an information-gathering probe to orbit our Sun and gather information on planets around our star is not new. The probe would scan the exosphere or habitability zone of our Sun, look for intelligence and then analyse fully the state of technological advancement of that intelligence , and examine closely all the attending planets and their makeup.

Oddly enough, cases are on record of our satellites being somehow interfered with, right from the initial launches. Their signals have faded or switched off for certain periods then mysteriously switched themselves on again. Our radio signals have bounced back from time to time as though hitting something 'out there'. During the space programme leading up to the Moon landing, objects were sighted in space with unfamiliar orbits. In other words, not put there by humans, which of course need not be UFO's, they could be chunks of rock that approached the Earth on a different track.

Perhaps it was an approaching alien probe that bounced a TV programme back to Earth that streamed into homes in the south of England. The programme had gone off the air in the USA years beforehand. Ships at sea have been known to receive an odd message or two using a maritime call sign that had long been discontinued, or the old-fashioned morse code signals of a series of dots and dashes it is a pity that we were not told what the message said, but no one knew how to interpret morse on the vessel.

To be sure, any probe from an alien source would have much to relay, on encountering the fertile Earth and its immense variation of life forms. The planet would be viewed as literally seething with life, including all the hostile and arid regions with the polar zones with the bears, penguins, Arctic foxes and so forth. There seems to be an awful lot of 'real estate' 'out there' of newly discovered worlds that could not possibly support life as we know it and would have been passed by.

The probe, in its travels, would have undoubtedly seen many lifeless crater strewn or cold gaseous worlds on its journey to Earth, if its visit here was just one of a series of reconnaissance sorties around the 'local group' of stars.

It could have a power source that eventually propelled it to enormous speeds by a gradual build up of velocity and similar deceleration times achieved, perhaps, by nuclear means.

It is quite possible that by now we ourselves could have reached the stage of technological advancement where our probes were looking over the planets of another star system. Ancient Hindu texts seem to indicate some knowledge of nuclear physics and the ancient Greeks show a past knowledge of the atomic structure of matter. Our typical human 'fear of the unknown' in times past and the destruction of written works and scrolls of priceless knowledge, have retarded our advancement enormously. The aforesaid Hindu texts not only suggest knowledge of atomic activity but speak of fearsome weapons and flying craft delivering them.

The suspicion and fear of knowledge that was not understood and the dispersal of learned thinkers and philosophers and the periodic decline of civilised advancement saw only a seemingly endless rise and fall of progress, particularly in those countries that had many threats around their borders, unlike island civilisations.

For all that, one might ask, what is the rush? Science tells us that our star still has four and a half billion years of life left in it. Surely, before that time comes we will have discovered all we wish to know and will have joined the abode of the 'creators' and be long gone from this system. This, of course, assumes that we have the will to eventually (implement) earthly protection system in order to save all our progress from celestial dangers in the form of comets and asteroids. If we do reach such an advancement, all the worthwhile minerals will be plundered out from, not only Earth, but from Mars, Venus, and no doubt all the sizeable asteroids. Our productivity for plundering destroying and denuding our forests, depleting our planetary resources seems just as strong as our gifted abilities to create, invent and nurture, in any case, star nurseries or stellar creation centres exist in abundance in the cosmos and surely attendant planets as well, therefore a multitude of worlds being created will far outstrip any operations to deplete them. Planets may not support human life are no doubt replete with rare minerals formed during their composition so no life forms would be harmed by future cosmic explorers plundering them.

We may have to depart our world, or at least some of us, at some point long in the future, the activity of the Sun may have become dangerous to us.

Our abandoned huge orbiting 'Freeman Dyson' biospheres and other excess population support facilities circling the exhausted shell of Earth, all run down cold and lifeless awaiting their fate, which will soon some upon them when the Sun flares up and roasts its inner retinue, or perhaps the entire system. By then, we will be long gone from our system and reside perhaps on another earth-like world.

If the universe is fifteen to twenty billion years old (and the exact age is far from certain), some civilisations must have already reached the aforesaid state of advancement and it is hard to imagine them still occupying a 'human type' vulnerable flesh and blood type body form, assuming they ever had one in the first place.

Perhaps such beings are now the creators, travelling the universe at will and looking at us in the same manner as we would termites building their air-conditioned mound. A brief passing interest, and with regard to those beings, surely, after such a timescale, no barriers would be imposed on them. So many 'Ensteinian' concepts have proved to be correct, that perhaps there is a speed of light barrier. However, at the start of the jet age we thought there was a sound barrier and aircraft broke up trying to penetrate it. However, penetrate it we did, in that particular case, with good design and improved power. For all that, some 'Ensteinian' concepts have now been challenged successfully. Nobody can be right all of the time.

However, when we start to consider moving beyond the speed of light. We hear about mass and start moving into the world of physics. The corresponding increase in mass needing infinite speed and so forth. However, as usual, we encounter something that seems to defy the laws. The C.E.R.N. project in Switzerland found that particles they studied seemed to travel faster than light, also, Dr John Allen and Godfrey Endean, two Oxford University scientists, announced that electromagnetic fields in the Crab Nebulae seemed to be travelling at twice the speed of light.

To be sure, there will be other intelligent beings existing on planets that have become close to mineralogical depletion and having massive population problems, but to a much worse degree than ours. Such beings may occupy the only habitable planet within the exosphere of their sun and have no convenient Mars-like or Venus-like worlds to terraform in order to solve their problems.

They would be looking for a new home. Which way to go? 'Upmarket' or 'downmarket'? Perhaps towards newcomers like ourselves out near the rim of the galaxy. They could no doubt impress us, but they would not be likely to impress the other way and why go to an older star. A young to middle-aged star, such as ours, would be far more preferable, particularly with three planets within the habitability zone for the taking. So … are they on the way here now?

Humanity apparently co-existed for a long, long time living among and existing like primates did, but our amazing brain that many see as a bequest from the 'cosmic creators' as well as evolutionary or spiritual in origin, would not let us rest. Already, this fantastic organ was streaking light years ahead of the apes and, with its massive over-endowments, was calculating, observing evaluating and creating and moving faster away from the simian types still foraging around the ant hills and bushes, that we may never have had any evolutionary connection with any way.

The intelligence gap between human and simian kind is now enormous. Even back in the twenties when basic radio was in its infancy, advancements proceeded in other branches of technology. We thought we were pretty smart even then, and we began to produce our encyclopaedias and volumes of pictoral knowledge and all the things we knew would soon come to pass.

But as well as a driving force within us to obtain more and more knowledge and technological advancement there is another force within us, a typically human one, that had many names, 'material gain, exploitation, greed, the hunger for a quick profit and personal enrichment', and the financial gain in creating radio and transmitting capability which is indirectly responsible for attracting (perhaps undesirable) alien entities towards us, invited by purposely revealing our solar system, out to any star systems with planets housing occupants of the aforesaid category wondering which way to go, that may exist within the vast distance now reached by our radiated emissions.   Those occupants may even have the ability with highly advanced equipment, to pick out even our weak radio and TV signals from the more high-powered transmissions.

The aforesaid urge to develop radio for commercial gain as far as we could was enormous, and the strong motivation of the above factors, precluded any possibility of anyone suggesting, "Wait a minute, let's sit down and talk this over.   Radio waves travel at the speed of light don't they? .... Well, only the other day I read a story about other world creatures possibly existing on Mars or on planets light years away. Eventually they would be aware of us here. They may be hostile. It could be like that war of the world's story. They might be looking for a new home and we are handing them the world on a plate".

"Don't you think we ought to consider a little – maybe think it over"?

"Yes .... mmm, I see what you mean. Perhaps we should concentrate on developing our receivers and listen out a little in the first instance".

Well we did not sit down and talk it over, that is for sure, but perhaps our hypothetical aliens did. Perhaps they are a bit smarter than us and, while we are chattering away so loudly, they are just quietly listening. They may have proceeded on a different tack by taking steps to ensure they could somehow blanket their radiation by some sort of planetary envelope to prevent them escaping and then developing their receiving equipment as refined as possible.

Now everyone within hearing distance knows we are here, but though we scan on many frequencies we have not heard so much as a peep out of anyone else. No one seems to be tying to talk to us, or are they.

Quite frequently, the ufologists claim that various Government Agencies are covering up the facts regarding UFOs. Has a 'D' notice been applied to a clear intelligent signal from a close extra terrestrial source? Would such agencies, worried about social disorientation and cultural shock be any less reticent with releasing information on an E.T. 'signal' source if they are going to withhold information on E.T. 'craft'? There would not be much point in explaining so vividly in all the media about the various sky searches and S.E.T.I. programmes only to be coy about announcing that they had found what they were looking for in the first place. It is fairly certain that, if any attempts at withholding such information were made, they would be very difficult to enforce. The excitement of such a discovery would be immense and there would almost certainly be a leak; someone would talk, unless of course, they knew they would be out of a job if their funding and broadcasting licence was revoked.

In any case, why, in this age of assumed enlightenment on things celestial, would the authorities have this fear of releasing such information to the masses? Admittedly, we took to the hills in 1938 with the famous radio broadcast of *The War of the Worlds* with Orson Welles convincingly telling the USA that the Martians had landed, but that was nearly 80 years ago. A spokesman for a radio telescope unit stated, "We would be delighted to tell the world if we detected an alien signal but he could be immediately silenced just as the radio station in 1947 was when broadcasting about the Rosswell incident. Their broadcasting licence was under threat of withdrawal and the transmission was effectively stopped.

There has been a deluge of data streaming into our senses from science fiction to science fact, including Mars, Venus and Moon probe landings, and in the case of the Moon, a dozen assorted astronauts bounding about, planetary round trips and so forth. Surely, by now we are sufficiently cosmically aware that other beings might just be doing the same thing? Surely now we should be mentally conditioned and mature enough to accept it? However, we must keep in mind our initial comment about armchair entertainment being quite different from reality.

It has been suggested that the various authorities have evidence of alien entities and actually have hardware and even body forms in secret locations, and are willing to foot the bill for S.E.T.I. programmes to maintain the ruse. It does seem strange that the funds seemed readily available for S.E.T.I. yet no cash was forthcoming for UFO search organisations, such as Ground Saucer Watch and others trying to get to the bottom of the UFO phenomena, and the whole thing was played down and shoved around like pass the parcel, with science not wishing to get to grips with it. This situation still prevails today, with cases on record that do seem to be credible, being filed away and ignored along with many examples of artefacts and anomalies that do not fit in with current theories, and historians are terrified of their orderly chronological explanations of human progress and advancement being stood on its head. Consequently, such oddities as a ribbed boot print in stone or the fossilised human brain found from the carboniferous period, or a machined cube found in a coal seam, are all quietly ignored. In other work, we have mentioned these 'ooparts' (out of place artifacts).

We can speculate on the possibility of the hypothetical alien race, being aware of allowing their radiation out into space and revealing their presence, but surely not all of them. Nevertheless, no other beings seem to be communicating, at least not in our direction. Perhaps we, in our remote outer region of the galaxy are seen as unlikely candidates to be in possession of the technology necessary and to receive and interpret their communications, and they see it as more likely to get a response by beaming their emissions in the other direction.

To be sure, it would be extremely exciting to receive an intelligent and obvious E.T. communication from space, but also extremely frustrating if it emanated from so far away that those who reply would be all dead before the senders would get to hear our message "Receiving you loud and clear".

When we speculate on the existence of other planets and with regard to their abundance, assuming most stars do have earth-like planets, is that for every star we can see with the naked eye there are 50 billion others we do not see. Many of them are multi-starred systems, binaries and triple arrangements like our nearest neighbour Alpha Centauri – although it has been suggested our Sun is part of the binary system, we have no real evidence for this.

It is calculated that extra terrestrial civilisations should average out at one every 40 light years, but this is based on them being evenly distributed around our part of the galaxy. If there is any substance in the calculations, it could still mean that quite a lot of alien eyes and ears are busily trying to decipher our broadcast and noting our area of the galaxy as worthy of a visit. These aforementioned tell tale radiations of ours could also assist any cruising cosmic craft to get a fix on us. Craft that may have been sailing the universe for generations with the original members that departed their world for whatever reasons, long dead, and their descendants operating the craft.

However, it does not seem likely that the 'chosen ones' of any hypothetical alien race that had to hurriedly vacate their planetary home, because perhaps a 'free world' was approaching it which would be too much even for their orbiting planetary defence missiles, would just blunder off randomly into space. There would be time enough to plan a logical course, and it would be reasonable enough to head for a single star not too far into its main sequence, and possible of similar mass and so forth to their own. If they did happen to stray into our flow of communications and turn their detectors our way, our star may just fit the bill and they would change course for Earth.

Clearly, from our point of view, if an E.T. intelligence detecting transmitted radiations from our system signalled to us this would be very exciting. We have found another Intelligence in space, and after we had adjusted mentally to the shock, we may begin to see the futility of asking questions and waiting say 30 years for a reply but we would feel quite safe about it, as it would be our successors who would have to deal with it.

However, any Intelligence, off Earth, that had the capability of fast interstellar travel, would rather utilise it for the obvious reasons of getting to the source of those transmissions as soon as possible, but also play safe by not revealing their own location and to have a good look at us first. Alternatively, they may decide, if having such advanced capability, to send a probe first to play even safer and to update the past information they had been detecting, which would already be as old as it took for such data to reach our world. Then on their arrival in our system, they would evaluate the data in their probe, in order to assess our current capabilities.

It is interesting to note here that in the early days of our space satellites some strange things happened. Scientists detected highly unusual signals emanating from a moving source in space on 14.286 megacycles, but the only satellites in orbit at the time were transmitting on frequencies of 20.005 and 40.001 megacycles.

Furthermore, when other satellites were up in orbit it appeared that 'something' was analysing them or checking them over. Some went dead and restarted again later. Consider some of the following occurrences; blinking lights on the Anna 'Firefly' Geodetic survey satellites were seen to fade and then stop altogether, then start up again much later, when they resumed flashing as normal. Telstar I and II both ceased transmitting, then being 'off air' so to speak, for long enough to assume permanent failure, they started transmitting again, first one, then the other.

A Soviet satellite launched in 1965 with an estimated life of 'decades' disappeared altogether. A US space official commented; "It is incredible; it would be contrary to the laws of science for a spacecraft in orbit to decay that fast, unless there were some means of propulsion aboard to change the orbit". We must remember however, that our knowledge of solar activity was less advanced at that time. It would be interesting to know if an outburst had occurred then, causing an inflation of the atmosphere.

Ordinarily of course, is that the satellite would have to have a forward facing rocket motor to produce a 'burn' to slow the forward motion of the satellite down enough for it to become slow enough to be dragged out of 'free fall' (or its orbit) to plunge into the Earth's atmosphere and burn up. The orbit of all satellites eventually decays because they are purposely close to the Earth's atmosphere to get the best use out of them, particularly if they are Earth survey satellites. Here, a picture is conjured up of an advanced probe giving no radar return, advancing on the hapless satellite, rather like something out of a James Bond film, with opening doors that eventually swallow up the satellite, but as said, an inflation of the atmosphere would retard the orbital speed of any satellites reached. Records of solar outbursts are kept - did one occur in 1965?

It was not only satellites that appeared to be subject to scrutiny. The power on the Venus bound Mariner II spacecraft was cut off, then later suddenly returned. Messages have bounced back to Earth long after they were sent, such as the aforementioned TV programme coming into homes in the south of England years after it had gone off the air in the USA.

To be sure, theorising about alien probes is not new. Early in 1973, a chap named Duncan Lunan decoded and analysed transmissions received in the early days of radio, and deduced that they were messages sent by the inhabitants of a planet circling the binary Epsilon Boötis 200 light years away. It is rather doubtful, in view of the distance that these messages if sent by an off Earth Intelligence, were meant for earthly ears.

When it was said that no one is talking to us, I mean that radio astronomers do not appear to be able to extract from the mass of signals they do receive from obvious sources (to them), such as pulsars, any signals of a purposely transmitted intelligent variety, which in their skill and experience they could separate from the mass of natural chatter of the aforementioned pulsars, close orbiting binaries and neutron stars, etc. For all that, however, consider the following rather startling statements: "We have been receiving radio signals from outer space .. They are not natural signals, but have been transmitted by civilised beings with sophisticated transmission equipment", Dr Nikolai Kardasheve, Radio Institute, Gorki, USSR. It was usual for such institutions to share such information in order to verify their discovery but perhaps the Russian Authorities prevented this.

Surely, there have been some modern day 'eyebrow raisers'? As long ago as October 1973, The Radio Institute at Gorki in Russia reported radio signals from outer space in bursts, lasting from two to ten minutes and noted "That their character, their consistent pattern and their regular transmission leaves us in no doubt that they are artificial in origin". Professor Kaplan of the Institute believed that most of the signals come from civilisations somewhere in our galaxy and are older and more advanced than our own. This sensational discovery should have been flashed around the world and would certainly have helped humanity to mentally adjust to it and prepared humans for future revelations.

A radio astronomer of Stanford University USA believes that somewhere out in space a message from the stars may be on its way to us in the electronic brain of a messenger probe.   Such a probe, now analysing Earth and its species, would surely see the human entity as the wildlife of the planet.

We put other creatures in zoos or eat them, we ravage our planetary environment in our ignorance of the rarity of life-giving worlds such as ours. We have a natural proclivity toward violence and have been killing each other before our recorded history began.

However, apart from the negative side, our obvious positive achievements and intelligence would be noted and our space travel ambitions, and the probe would have reported on the suitability of Earth for adaption by other life forms. The probe may coldly and logically suggest that we are a dangerous species and, since we appear bent on self-annihilation, it might suggest that in this regard they could help us on our way.   Some speculate that the demise of the dinosaurs was brought about in this manner by our cosmic ancestors preparing the world eventually for us by eliminating the undesirable occupants.   The danger is that we may now be seen as the undesirable occupants.

Having once detected our emissions, it may be necessary for an alien race in fairly close proximity to send a probe to listen in to our weaker transmissions. But the far stronger ones, such as those pumped out by radar defence organisations with the ICBM and early warning facilities and so forth, are in the megawatts range, not just kilowatts, and certainly the radiations from the big dishes we use, such as Arecibo in Puerto Rico, would be streaming in loud and clear to any hypothetical alien races within range of their 'voices'.

If we consider the 1920's as being the onset of radiation output, such emissions could now, it is estimated, have probably reached something like a hundred sun like star systems, so it is perfectly possible that at least one civilisation is now listening to the sounds of Earth.

If they have the technology to listen, such a technology would be at least equivalent to our own, possibly more advanced. If, in their prehistory, they have known war and conflict, it may be that unlike us they have learned something from such negative actions and perhaps risen above such things. In this regard they may well have, as previously said, found a way to suppress their own emissions while sifting carefully through ours and, eventually they would know that we still pursued such pastimes as confrontation and territory disputes and were occupied in conflict, which would be obvious with our war bulletins from somewhere on the planet, assuming they could detect the weaker transmissions.

Even as long ago as 45 years, a Stanford University report stated (in 1971) "We conclude that if we continue to broadcast our TV signals for another century, Earth will be detectable up to something of the order of 100 light years, which could announce our existence to beings on any of the thousand or more likely star systems within that range.

It can be seen from the above, that a warning, or wary approach to radio astronomers, was being advocated, who were actively engaged in purposeful transmission, by blasting out intelligent radiation, and hence our location in the galaxy, out into space.

Unless the plans have been shelved for good, one assumes that we will eventually have a Moon base. At one time, there was much discussion on this concept and actual plans drawn up by NASA. Now we are doing the same thing for a Martian base. Even a simulated Martian environment was constructed, a facility where the occupants were supposed to stay inside, no matter what and be entirely self-sufficient by growing all their own food and, one supposed, recycling all their water, but as far as I can recall, as soon as one occupant took ill, they 'phoned for an ambulance.

If a signalling dish was set up on the Moon along with the construction of our Moon base, it has been stated that even a modest sixty watt transmitter based there would produce an adequate signal with an aerial dish of less than twenty meters that would enable us to have direct communication with a planet in a system some twenty light years away. However, a computerised satellite can do the job quite adequately from Earth orbit, but the signalling dish would not be the only reason for a Moon base. Yet the whole concept of any further lunar interest seemed to cease after Apollo 17. Of course, with the space shuttle development the question of expense and a 'been there, seen it, done it' attitude prevailed.

It has been seriously suggested that an alien presence was found on the Moon, which was the reason for the termination of lunar activity, by NASA. In my *When the Moon Came*, I mention various communications that came into our lunar excursion modules. The things astronauts had seen and comments by various NASA employees such as, "There are alien constructions on the Moon waiting to be discovered". However, claims have been made to suggest Apollo 11 encountered aliens, yet six more missions there took place.

To be sure, many hypothetical intelligences have been hearing us long enough to have arrived here or, at least sent an exploratory probe in order to beam communications picked up from Earth back to its own planet in some fashion. Perhaps the memory banks on some alien world are replete with information of all descriptions about planet Earth and, if the alien occupants are not already here, perhaps they soon will be.

Their designated landing areas may already have been decided upon and plans formulated neatly setting out their goals and objectives.   They would know the geography of Earth and its seasons, the temperate and hostile regions. They would be familiar with all our defence systems and their locations, all our armaments and the weapons technology, the capabilities of the various nations, and most importantly, analysed the atmosphere and the presence of any bacterial and microbes present in Earth space that may be detrimental to them.

It is perfectly possible of course that the alien presence, alleged to be in our skies are well aware of the true nature of human beginnings and any 'cultural shock' we may be about to receive would include these revelations to us. Astoundingly 'they' may even have (at least their predecessors) been responsible for our own origins.

When we watch the quality programmes on our television sets of the advancement of science, in the realm of the physicists and geneticists, it becomes obvious that in many cases they sometimes astound themselves in just what they are discovering, and what it seems likely that their descendants in the near future will be doing. Postulations today will become factual achievements tomorrow.

They also make it plain that so many other of the sciences are beginning to lag somewhat behind in comparison, such as medical science, which (though obviously advancing) still has not solved the problems of cancer, aids, muscle wasting diseases, and so forth, and with regard to the nuclear sciences, we have created the monster but we cannot kill it, and so bury it alive to perhaps escape and become a dangerous legacy to our descendants. Burying it is like sweeping dirt under the carpet.

There are many things that possible advanced alien entities, revealing themselves to human kind could perhaps help us with and solve many of our problems. With our present knowledge of metals, even titanium, would melt before it reached the sun if we tried to launch it there (our nuclear waste) and it would blow back on the solar wind.

Even if prolonged observation of the Earth has not taken place and other world entities may not have witnessed our early history, they may know quite a bit about us from the decades onwards since the advent of our radio broadcasting, due to the fact we have told them we are here and sent much uninvited information about planet Earth toward their world. If such entities are clever enough to detect our weaker transmissions, they will have received programmes on every topic imaginable since our television transmissions commenced.

They may even know the political leaders, the military leaders and their ambitions simply by analysing our languages and the TV and radio broadcasts of all descriptions. 'They' may know more about the Earth than the average 'Earthlings' knows.

Although the hypothetical aliens may have matured in their behavioural traits with regard to actually going to war with each other, it would be almost certain that they would possess adequate weaponry of some sort for at least defence purposes, to protect the planet from wandering celestial materials, and if they were desperate for a new home, for whatever reason, they may decide to bluff us by making a London-sized hole in Australia's central region with a simple message 'Look, this is what we are capable of, don't bother us and we won't bother you'. After all, for all our worries about the increase in our numbers, it is estimated that the entire world's peoples could fit on the Isle of Wight (of course, they would all have to remain standing), but it serves to make the point that if the aliens had made the same calculation, they may feel that there was plenty of 'Lebensraum' on Earth for the refugee band. After all, they could not ship their entire population here ... could they?

If the aliens did adopt such a plan, it would still be a very risky business for them. If they chose this kind of action rather than ruthlessly eliminating humankind, they may, in the long run, wish they had done so. They could never expect to live in peace from us in some kind of splendid isolation. There would always have been the curiosity seekers, even some hoping to capture an alien or two for a travelling show or just to prove they could do it. We would never leave them in peace. They would be on a reservation of their own choosing. Humans would come along for a little trading hoping for rare alien artefacts. There would be problems of alien bacteriological strains unknown to us. The possibility of Earthly strains we have long become immune to, wiping them out. Their numbers would have to be strictly controlled. We would worry about them possibly wanting more territory.

We would wonder what they were doing in their science, whether they would be plotting to overthrow the Earth at some future date. Eventually, some high-ranking military personage would state in secret sessions on the 'alien question' – "Better to destroy them now than wait for them to attack us. One bomb would do it". There are always plenty of high-ranking 'Hawks' in the military that think in this fashion.

It has been said, that during the cold war, this was actually suggested as a way of removing the threat, even accepting the retaliation casualties that would be bound to occur after a surprise attack knocking out most of the opposition's forces. These hawkish proposals would arise in the insane logic of living under the threat constantly of nuclear war, and probably instigated the stated and avowed policy of the west, never attacking first in such a scenario.   This statement has the two-pronged effect of stopping the opposition (hopefully) from thinking, or acting, on similar lines and of relieving the mental pressure on ourselves that, having made such a statement, the opposition would abandon *their* considerations in this regard, but that of course would always be a gamble.

Although the Earth is 70% water, we still have adequate landmasses and, if agricultural science can solve the feeding problem by encouraging hostile areas to produce foodstuffs, which would go hand in hand with weather control facilities, we should be able to adequately feed the world masses. Nevertheless, some form of population control may still be necessary. We can resolve all our problems.   All we need is the will to do so. An eastern quotation states 'The difference between what is possible and impossible is the measure of one's will'. Why should droughts ever exist on a world with seventy percent water? Desalination plants existed over 50 years ago. We can pipe oil, why not water?

Currently it would appear that problems are being encountered, particularly in third world countries of the consequences of ignoring the dire warnings of the fact that the world could become unable to feed its masses if they do not self-regulate their population expansion. To be sure, we could all get on to the Isle of Wight as long as we were happy with just a square meter apiece.

To return to our assumed alien star ships, their computers would be programmed to search for stars similar to their own of a stable nature, with at least four or five billion years of life left in them, and ours would fit the bill very nicely.   They would only have to come within 90 light years before encountering our obviously intelligent transmissions, and then a quick course change towards Earth would be all that was necessary, and after beholding such a sight as Earth hanging in the heavens after all that wondering in the void, perhaps only encountering crater strewn or gaseous worlds, they would be very reluctant to pass us by in the hope of finding something better, or because Earth just happened to be occupied. If they had advanced to the points, we speculate upon such as traversing wormholes or warping space/time, such distances from Earth would not be a problem.

Because they have bridged interstellar distances, it would be easy to assume, due to this very fact, that they have enormous intellect, but would this necessarily be the case? We all take our electronics and computer technology for granted, but how many of us could design a computer or even service one. Most of us could not even repair a simple radio and outlets that undertook such repair work on them and other electronic equipment are fast disappearing, with the advance of modern electronics and miniaturisation, smart phones etc.

If the average man in the street was washed up on a desert island he would be like a time traveller lost in the Stone Age, and only be able to describe the wonders he previously took for granted all around him. The hypothetical alien travellers may have depended on a super brain or computer in their star-ships to accomplish everything for them to heat, feed, monitor and administer to their every whim, and though such beings may have originated in a supremely advanced environment, they would be quickly reduced to a very basic state on arrival, and have to begin the long return to the re-inventing and creative mode, on some hospitable world.

Nevertheless, they should have a naturally advanced intellect and their onboard central brain would have computed many different scenarios for them to expect if they encountered any technologies having emerged on planets they would encounter and, with regard to Earth and its occupants it may only be necessary to draw the correct plan to deal with us.

It is entirely possible that alien entities, perhaps an armada of them in huge stars-ships, are now approaching our system and that the occupants of the craft have never themselves experienced a planetary existence and know only of such things due to the onboard facilities for education of the offspring having lived and died over many generations that were set up by the original beings who long ago departed their star system, and had discharged the corpses of those who had passed away, into the void, ahead of their star ships and eventually stream into our atmosphere as meteors before their main craft arrived.

The stories would be related with scenes of their former world and the reason for their hasty departure, perhaps a celestial threat of an approaching planet being dragged along (after possible capture) by a long orbiting binary star companion to their own sun, or even their sun itself becoming erratic or unstable. This scenario inspired the early science fiction film *When Worlds Collide*. The star passed by but one of its planets collided with the Earth. The escapees blasted off to rendezvous with a favourable planet orbiting the new sun.

Some writers have suggested that our ancestors could have arrived here on Earth in the very same manner long into the past. This would certainly explain the curious phenomena of the human brain that appears to have needed much more evolutionary time than is ascribed to its development to produce such an 'over endowed' organ. Certainly, any race of beings arriving in such a manner would be bound to degenerate, just as the example of the children in the *Lord of the Flies* suggested, and like children begin again in the learning mode.

Many factors would be responsible for such degeneration, not least the exigencies of survival, hostile planetary conditions, the losing of all roots and knowledge of pre-history and a gradual belief that they were indigenous to the planet, but still being the cleverest species with dominion over all others. The human entity may be so old that we have planet hopped more than once since initial evolvement in some remote system, or possibly even from Mars or Venus, having caused a massive 'greenhouse' effect there, which we seem intent on repeating with regard to our own world.

Such a hypothesis would explain the extremely ancient boot prints showing a patterned sole and stitching from a time before human existence was thought to have begun.

They may view the Earth as some form of 'promised land' their 'brain' had spoken of that they would inevitably reach and, having viewed the vast unpopulated areas on Earth, begin their skilful bargaining process for terms of settlement, perhaps by the promise of enhancing our own advancement. Nevertheless, they would, in this regard, be a bit like a human colony setting itself up in lion country and expecting the lions to go away and leave them alone. They would never find peace among humans, once we found they were vulnerable.

When viewed against the backdrop of the estimated age of the Earth, human presence on the surface of this world has been like a blink of any eye, especially if compared to cosmic time scales, and when one considers the cosmic threats that could annihilate us with cold detached impartiality, it is a wonder that we have survived for as long as we have, and perhaps the comet that came into our system and was torn apart by Jupiter finally taught us a lesson that should not be ignored. When we analysed the severity and size of the impact of the torn apart remnants of that cosmic visitor in Jupiter, it gave us a clear idea of what we could expect if we were on the receiving end. Major annihilation events have occurred in the past, so we may be a lot nearer to the next than we are far from the last.

While we are aware of some periodic comets, it brought home the reality of unknown visitors arriving at any time. Furthermore, we also have the orbital asteroid threat and, while we have named and logged the orbits of the main bodies, there are plenty of large ones left in the asteroid belt waiting perhaps, to be disturbed on to an earthly course by some passing comet, and we certainly have not detected all of them.

It is hard to imagine other assumed planetary systems having any meteors or asteroids unless that system had an 'asteroid belt'. Indeed, it is also hard to imagine that the asteroid belt was not once a planet itself; after all, it is where a planet should be. Astrophysicists point out that there is not enough material there to form a planet but how much material could have escaped the orbital path and headed into the sun, the Earth, now obliterated by erosion, of the Moon or even the other way toward Mars? And into the earthly oceans.

It seems certain that almost every little stone and dust grain that streaks into our atmosphere once existed there as part of a whole body, disturbed by the grinding action and collisions causing changes of orbital direction. If the asteroid belt was once a planet, perhaps the main bulk of it has produced the severely pockmarked appearance of Mercury, as the planetary remnants streaked toward the Sun to be consumed, and of course burnt up as dust when encountering Earth's atmosphere the craters on Earth may be much larger if a good portion of the material had not burnt up are regularly discovering which were not part of the calculation. Of course, some seabed areas were once surface areas and vice versa, with marine evidence being found in our highest mountains. Such Earthly impacts must have been quite catastrophic, particularly if they hit the sea rather than a land mass, with the widespread flooding and tidal wave effects. As we have said, we must be nearer to the next that we are from the last and with the short time, human kind has resided on Earth, we may well have missed all the previous main events. When we consider the size of the Pacific, it must have swallowed a lot of the material from the asteroid belt.

If we accept the possibility of life on worlds elsewhere in space, many must be far in advance of us, and many must have lived and died in some catastrophe such as a nearby supernova, before life even appeared on Earth. Any advanced alien race that detected the possibility of a nearby star becoming unstable, then this would be reason enough for their formulating plans for an exodus for at least some of their race. It is possible to assume many wandering 'arks' of extra terrestrial beings traversing space looking for a home. It is fortunate for us that the stars in our neighbourhood seem to be fairly stable. We know that there are colliding galaxies in space – surely, some 'occupied' worlds must be affected? However, the possibility of other extra terrestrial races having evolved long before us can only be based on our original computations of the so-called 'Big Bang' occurring up to twenty billion years ago. But today, this figure seems to be in a downward mode and estimations of 11 billion years, and even less, have been suggested. In that case, perhaps none are more advanced than we are.

Tantalising evidence suggesting organic life has been found in cosmic rubble on Earth that no doubt originated in the asteroid belt. Comments such as "Where this originated something lived", have been made, and almost certainly these chunks of rock could never have 'accreted' under their own infinitesimal gravitational attraction, and as said, geochemists have stated that they must, at one time, have been part of the parent body.

It would appear finally, that at last some representatives of the powers that be, are at least talking about the orbiting cosmic dangers to the Earth and what we might be able to do about it. Of course, as ever, cost and responsibility factors are immediately raised, fortunately finance does not now appear to be affecting these studies and today consideration for how to deal with the threatening debris are being formulated.

To be sure, the time is long overdue when serious consideration ought to be given on putting up at least some form of protection with a multi-national agreement on the sharing of the technology, expense and contribution to the maintenance of such equipment and, as all nations of the Earth will benefit from such protection, all nations of the Earth would be obliged to contribute in direct proportion to their resources. However, security measures would have to exist to prevent their use in earthly conflicts built into their design.

To return to the theme of a possible extra terrestrial threat (as opposed to a random cosmic one) perhaps invited here by our 'vacancies' sign beaming into space, the possibility of such an orbiting powerful weapon defence system being turned around to point at us as a threat to achieve their aims of annihilating us or simply to dominate our world cannot be ignored, therefore a failsafe mechanism must be implemented in their structure.

Very careful design in such a powerful orbiting arrangement would have to be employed. For instance, a built in destruct capability, along with the impossibility for such weapons to detonate unless moving away from a large gravitational body toward a smaller one, and their inability to function the other way around.

Of course, much of the time after the 'Big Bang' was taken up in the process of forming conditions for life to exist. Initially, all was hydrogen (and, indeed, most of the matter in the Universe still is). The first generation stars probably had no planets orbiting them except perhaps a few gaseous bodies, more like failed stars themselves, as it was necessary for the death of the first stars in explosive demise to fling out all the necessary material for the next generation, and so forth, until there was sufficient heavy material to form planets and living things, such as we ourselves, essentially composed of material once cooked up inside the hot, fiery interior of a star.

If we take Epsilon Eridani as the hypothetical source of extra terrestrial intelligence, simply because this was the star that caused a bit of excitement in the very early days of Sky Search and S.E.T.I. programmes, a planet in this system, being only a little over ten light years away, would only need to be advanced enough to have discovered a suitable means of power for its star-ships to enable them to reach the Earth system comparatively easily. A fusion type power source could gradually power their interstellar craft over a long period; say about a year, up to an enormous velocity. If they have a much longer life span than us, their journey may not seem too tedious to bear, particularly with deep freeze capability possibly being utilised. After all, the mariners of old spent a good portion of their lives at sea during the early age of discovery and their lifespan was considerably less than modern humans.

The rigorous conditions and lifestyle of our forebears meant a shorter lifespan. Today, with our creature comforts and technology and medical discoveries, we live longer, therefore, the equation seems to be, the greater the advancement, the longer the life. 'Epsilon Eridanians' facing an eleven or twelve year voyage to our system may not be too daunted by the prospect of such a journey and may be looked upon as said in the same way as our ancient seafarers and explorers did on Earth, particularly if the Eridanians lived some 200 to 300 years as humans will surely do in the future.

So … was it Epsilon Eridanians that were appearing to all those biblical characters in their 'flying shields', pillars of fire, rumblings and roaring's and figures with calf's feet (rather like astronauts boots) and suits of burnished bronze? Regarding the biblical characters, it would appear when seen in modern day terms, that their descriptions were relating to some form of extra terrestrial activity on Earth that impressed the observers enough to write so descriptively about them and for their writings to last through time.

All the above means, naturally, that we were being well surveyed and looked over long before our telltale radiations drifted past their star system. If this scenario is true, then unfortunately it does not say a great deal for human behaviour patterns to be observed for millennia and still the observers are quite reticent to initiate any form of contact with humankind, which should not really surprise us.

When one considers the comparatively short time we have utilised any form of radio receiving facility, intelligent signals may have come our way in the past and now the senders have moved on to the new quadrant in the cosmos, having received no reaction from our particular sector, or they may view our area of the galaxy as a 'cosmic backwoods'.

We may speculate on *alien* entities perhaps arriving here after sailing through space in a cosmic ark, but it is interesting to conjecture whether any plan has been loosely formulated for a human 'exodus' of a chosen hypothetical group to leave Earth for another star system in the event of some cosmic threat, together with carefully considered artefacts to accompany them, and an interesting exercise for the education authorities would be to have the students in various educational outlets complete a list (simply to make them think about it as a mental exercise, or just for fun) of what books or artefacts they would consider essential or necessary for a chosen group to take along with them on a 'one way' voyage, bearing in mind that space in the craft would be at an absolute premium (assuming the cosmic 'ark' does reach an earth like world, there is every reason to assume metallic ores would exist there so the pilgrims may eliminate the stone age and commence their advancement from the possible mining out of such ores).   Then go on to list what professions the chosen ones (in equal numbers of gender) would, or should be, proficient in, on the assumption that they could reach, with their own power sources, our nearest star system Proxima Centauri and, if encountering a suitable planet, start a new life there and ensure the continuance of the human race.

It would not be at all surprising that a volunteer crew could be found today for such a venture, even if they knew it would only be their offspring who would encounter the new world, and even knowing that it would be a one-way ticket to survival or oblivion, there would certainly be no turning back if the Earth was in jeopardy.

It is interesting to consider just what protection such entities arriving in earth space would have in a legal sense from being mercilessly exploited, robbed or even murdered by Earthlings.

Are there any laws that have been drawn up to protect aliens? They would indeed be like the equivalent of St Brendan's monks, being allowed to exist only by kind permission of the natives.

In reality, they would be immediately surrounded by suited figures, who would whisk them away to some remote location, and the general public would quite firmly be discouraged from getting anywhere near them. They would be poked, probed and tested, closely examined, X-rayed, have their 'blood' examined and any alien bacteriological strains would be searched for at the same time as attempting to give them maximum protection from our strains. This procedure would be essential and anyone, or any area, they had contacted would mean certain human beings being placed in immediate isolation for similar tests, and the area of their Earth contact cordoned off and examined. They may carry a virus they are long immune to but which could decimate Earthly populations in a few weeks.

These precautions would have to be taken by any visiting aliens that may right now be in Earth vicinity, and any advance probe would quite likely be equipped with an expendable 'sond' to come into our atmosphere and carry out such essential checks, relay its information, then probably self-destruct.

The saga of the L.D.E.s or, long delayed echoes experienced in the early days of our radio emissions, have not been entirely resolved to every scientist's satisfaction. Some quite serious radio experiments were going on even in the late twenties, and it was found that signals sent out took precisely $1/7^{th}$ of a second to circuit the Earth and return to its source.

Other signal return times are known, such as how long it takes for one to bounce back from the ionosphere, or even from the Moon, which is why even before more accurate laser measuring processes, it was possible to be fairly precise about the distance of the Moon. However, it is when signal return times start to vary that something strange would appear to be going on.

One experiment in the twenties sent out signals of three short dots in rapid succession at intervals of thirty seconds, and when the echoes returned they were on the same frequency (as the pitch and tone were the same), but the intervals varied between the echoes, and it was speculated that if the distance between the delays could have been analysed, it might have built up a radio picture of a message. The process in reverse is envisaged and has now possibly been carried out in sending such a message or radio picture out into space describing Earth life.

Information or messages we might wish to send could be quite easily composed using simple dots and dashes and, if these were repeated across a moving sheet of paper, the varying dots and dashes would eventually build up a picture rather akin to one of those colour blindness test charts. If one looked at it in a certain way, perhaps by ignoring the dashed, the dots could form a definite shape, the shape of the human form, for example.

It is possible that just such a pictogram could be, even now, streaming into our receiving apparatus if we happened to be tuned to that particular emission line. Of course, the obvious emission lines have been looked at for the possibility of such things, particularly that of hydrogen, the most common substance, and therefore in human logic (and why not E.T. logic), the most likely frequency to choose, and this comes in at 1420.4 megacycles. However, when one considers the sheer volume of possible transmission sources, this is where the problems arise.

The L.D.E.'s were a bit of a puzzler up to the Second World War, but then obviously there were more important things to worry about. Mankind's seemingly favourite pastime was calling again and everyone was going off to war, and after the war there was a tremendous increase in radio frequencies and it was obviously a lot more difficult to pick out specific radio noises or alleged signals than it had been in the late twenties and thirties, but the tempting postulation still exists, as to whether they were returned by an exploratory probe traversing the heavens. If there was a probe it would have had much to occupy it from the six years after 1939 and indeed, the many wars afterwards.

The advancement and power output capabilities today are enormous and a lot of answers are in regarding things that puzzled and intrigued us in the past. Nevertheless, there always seem something new over the horizon to discover and study.

Amazingly, it has been suggested that extra terrestrial intelligences actually move among us today, or at least very humanlike genetic entities, and these have been 'produced' from abducted people unknowingly supplying their genetic material. Clearly, if somebody did know this on their own, with no confirmation or witness to back up their statements, it would be a terrible dilemma for them. They would find it far too mentally stressful to keep such a thing to themselves, yet if they did tell anyone else say, for instance, their close friends, they would quickly find those close friends becoming quite distant friends and, in the worst scenario, may find themselves certified. Whereas one would have little hesitation in agreeing that a person who thought he was Napoleon Bonaparte was destined for some form of 'certification', the possibility must be considered that this method of infiltration is one that E.T.'s may well consider for all the above reasons. They would be well aware of the derision that such a person revealing information about and being he suspected (for good reason, of course) as being 'not of this world' would receive.

One could imagine such a 'hybrid' entity thinking he was alone and communicating to a holographic image of his alien control, who then 'dissimulates' before the observer's eyes; what would the observer do about it? Clearly, such hybrid entities would be on Earth for a specific purpose and would, with their special powers such as perhaps telepathic mind control, be able to reach (perhaps as part of their mission) very high positions in all their persuasions.   Since alleged abductees have been listened to, and even hypothetically retrogressed, at least this treatment, in addition to lie detector tests, should be offered to anyone in the above category.

Special powers seem to be a prerogative of a few gifted individuals on this Earth and quite astounding feats, under laboratory testing conditions, have been performed, especially in the Soviet Union. One wonders if we all had them in the past, or whether they are an incursion into the still developing and largely unused neurones within the brain. One or two quite clever people have offered the suggestion that extra terrestrials have, unbeknown to us, brought our civilisation and technology forward quite considerably during the current century.

Consider the following: Dr Hermann Oberth, rocket pioneer and space authority, one time head of the US Cal-Tech Laboratories said, "We cannot take all the credit for our record advancements in certain scientific fields alone; we have been 'helped'". When asked by whom he replied "The people of other worlds".

Certainly, there are at least a handful of people still alive, whose grandparents may have witnessed the Wright brothers' first historical flight and now realise the great burst of technology that saw in the jet age, the space age, the Lunar landings and the Space Shuttle flights, and plans now being laid to go to Mars, all coming to pass in such a comparatively short time.

It is also quite astounding to think that the historic flight, previously referred to, could have taken place inside the shell of our larger transport aircraft, such as the Airbus industries' 'Beluga', a huge bulbous whale shaped aircraft, designed to carry portions of the Airbus to the assembly finish line.

Now there is serious discussion and propositions for 1,000 seater passenger aircraft. Although people in the future will witness great technological advancements, it is doubtful that they will experience anything to match that which a hundred year old person has seen in their lifetime. Many great leaps forward were made but, for all that, none to match the biggest one of all, achieved in July 1969 on the Lunar surface by the crew of Apollo 11, most certainly a date to remember, and a tribute to the finer achievements of human intellect.

However, (conversely) no greatly increasing time span seems to be making itself apparent between the times that we go off to war and begin slaughtering each other once again, and this activity negates and somewhat sullies our achievements in the positive mode making many people not inordinately philosophical to ponder whether the human race is doomed to extinction by its own actions. Although we are aware that we are causing ecological damage to the planet, we still pursue the actions that cause such damage. Although we are aware that war is a stupid, ridiculous, negative waste of manpower, human life and Earthly resources, we still pursue such activity somewhere on the globe. Perhaps we should, for the moment, put ourselves in the position of the alleged earthly observers many people are convinced are studying the world and its occupants, and ask ourselves how we would sum up the earthly activities of the races of this world from an alien point of view. If they had left such negative pastimes as going off to war with each other, hundreds, maybe thousands of years ago, would they be keen to contact the worldly inhabitants? I think not. What they would do, if having the ability, would be to remove specimens from the surface in an abduction programme to see what made them 'tick', so to speak, and a genetic examination and mental analysis would be carried out.

Most certainly, even if they had never travelled here but had good detection capability, they would have a pretty good compilation of human activities and facts about Earth in their memory banks that could be leisurely studied in their hypothetical radio receiving facilities.

When such entities had finished their studies of the wildlife of Earth (namely us) they would move onto the animal species that seemed so calm, placid and untroubled and lived so peacefully in family groups, only killing perhaps to survive or feed their young, rather than for the fun of it, as in the mentally disturbed or with organised massed killing units we call armies. They may see the grazing animals and those free of captivity as being the true peace loving occupants of planet Earth. From a distance of ten plus light years such beings would have had 80 years in which to study planet Earth by sifting through all the various emissions. They may even be able to reproduce our TV programmes and would know, from the quality programmes, the complete geography of life forms of Earth, the conveniently positioned Moon with once face providing an excellent observation platform, should they wish to come here themselves.

They would know the disposition of the continents, our weather systems and political and military structure and weaponry, our economic structure, and all that would be needed prior to setting off for our system themselves would be the sending of a probe in order to beam back, towards their advancing craft, enough data to bring their memory banks up to date and to enable them to complete their plans of action when arriving in Earth Space, which could, of course, happen anytime.

For all that, our radio astronomers still patiently search the cosmos scanning all those emission lines and nobody seems to be talking, at least not in the 'radio' language we understand. Some radio astronomers assume that there may well be intelligent signals among all the background radiations and that we may simply not be recognising them.

They would see tribesmen in certain parts of the world, living largely as they did in the Stone Age and other nations planning voyages to other worlds. They would see crude dwellings and pyramids on the same planet. To be sure the latter would interest them greatly and they may find it difficult to assign them to human endeavour. If alien entities do exist in Earth space they would most certainly have found them extremely interesting and may have found the Buried knowledge and details of Earth's prehistory that some suggest could have been buried there awaiting discovery. UFO sightings have occurred in their vicinity. It has been proved that a form of electrical energy is being emitted from the apex.

The ancient Egyptian priests were almost certainly the keepers and protectors of wisdom and quite probably did bury knowledge for future generations in some cleverly concealed place, perhaps in a golden casket, to await discovery in more enlightened times. It would have been considered that the lower regions would be a most unsuitable place to position such data. Those ancient priests would, in their day, be rather like the select group of physicists working in some cyclotron and talking in a language no-one else understands as they are seemingly entirely immersed in a world of sub-atomic particles, quarks, muons, quantum physics and so forth, and can only converse in regard to their work, with a limited number of people. The ancient Greeks, as clever as they were, were likened to children by the Egyptian priests. Solon who had many audiences with them, was told the story of Atlantis and the priests hinted at the great age of mankind with many such catastrophes having befallen the Earth over many millennia. One wonders (if the Atlantis story was true) whether the survivors would have influenced the Egyptians and South American cultures. It seems unlikely that similarities in the structures and certain items unearthed, would exist and arise simultaneously across such a wide expanse of ocean, unless some large land mass existed that made it possible for the assumed technological advancement there, to easily reach and influence them, in any case there are pyramid like structures all over the globe.

The circumstantial evidence for the 'Amerindian' nations originating from this mythical lost island of Atlantis is fairly strong. The common belief is that these decidedly Asian-looking people from the far north (the Eskimos) to the tip of Chile, trekked across the Bering Straits into Alaska and down into Canada and America. The problem is that their own legends tell an entirely different story. They speak of 'the land of morning' i.e. the east, as the ancestral home for all the tribes. Although the Eskimo people are decidedly Asian looking, the Eskimo legends have their ancestors arriving in great 'metal birds'. As for the rest, they all seem to have legends referring to their origins toward the east or into the Atlantic Ocean. Specific mention is made to Aztlán, and this includes the Aztecs, Toltec's and the Maya.

The Spanish did make some attempts to document or enquire of their origins and recorded that, "The natives seem to believe their ancestors came from across the sea". It also seems to be fact that the Conquistadors' landings were made easier than they possibly might have been, due to the strange coincidence of their arrival, seeming to fulfil certain prophesies for that time among the native Indians of their fair-skinned 'teachers' returning, as promised, to them.

With regard to the more northerly inhabitants, certain tribes of the now United States of America share the same belief. Sioux legend states, "The tribes were once united and all dwelled together on an island". The Iowa tribe also state, "At first all men lived on an island where the day star is born". With regard to the aforementioned coincidence of Spanish arrival and a prophesy, the twelfth Inca, Huaya Kapac, stated that, "During the reign of the thirteenth Inca, white men would come from the east".

Hermando Cortes was greeted by the Emperor Montezuma in AD 1519 as the re-incarnation of Quetslcoatl, the fair skinned 'God' of the Aztecs who had brought civilisation to them from the east but had left, promising someday to return. What was the influencing factor behind the mysterious Cahokian culture that flourished around 1200 years ago along the US Mississippi River? They had fine artwork and intricate carvings and built strange pyramid shaped mounds described as ceremonial; they also had sophisticated knowledge of an astronomical nature.

A skeleton of a being was found, that had been laid out on robes made of 20,000 pearls.

This civilisation existed near Collinsville in Illinois USA and it has been estimated that around quarter million inhabitants occupied the site.

The circumstantial evidence for all the Amerindian peoples originating from the east is quite compelling. The similarities of cultures on both sides of the Atlantic Ocean is compelling evidence for a once existing centrally placed original source, and the mythological Atlantis certainly seems to have been that place.

Francis Hitching, in his *World Atlas of Mysteries* mentions the strong similarities to 'Egyptian' peoples that historians state are among the influx of the various races and foreign peoples that the evidence suggests made up the Amerindian nations. Sculpted heads depicting bearded entities of quite 'Semitic' appearance from 1500 BC are also mentioned. The Native American Indians are all quite beardless by nature.

It is often said that if there is an extra terrestrial intelligence in our airspace, then they would show great interest in the pyramidal constructions of Egypt and, indeed, the other mystifying Tiahuanaco and other South American wonders, that so perplex us in their purpose and methods of construction. This is of course assuming that such entities are not already aware of them. They may already have all the answers, or, as some suggest, their processors were responsible for them.

However, with regard to the best of the Egyptian constructions, we have seen in recent memory our own modern constructions collapse like a pack of cards in areas such as Great Britain, with little or no seismic activity to blame it on, and have to admit sheer bad design and human failings. A good example of which was the Ronan Point disaster a series of flats occupied a corner of an apartment block and totally collapsed.

The event occurred due to a lower unit on the corner of the block behind destroyed by a gas explosion and, with the weakening of one unit those above it all fell down one after the other. It was sheer good fortune that it happened when few of the occupants were in residence, sheer bad design, they perhaps should have, before going to the drawing board, consulted American know-how who have great experience in building such lofty structures.

In other parts of the world we have seen our bridges collapse after a moderate earthquake, when impressive buildings still stand that were built over two thousand years ago (in Greece, for example), an area quite prone to seismic activity (but we must pay tribute to the knowledgeable Victorian engineers in building bridges, ships and railways). Yet they and the fantastic Egyptian structures still stand as a testimony to obviously quite superior skills and advanced knowledge of geometry, construction, engineering, mathematics and weight distribution, and the precise fitting and laying of huge stone blocks without any of the huge impressive yellow painted mechanical wonders we see around today.

Perhaps we would benefit greatly from the knowledge said to be buried in the Khufu Pyramids of Giza, constructed 5,000 years ago.

We are all familiar with the typical 'tourist brochure' image of Egypt, with the setting of the sun and an Arab or two passing by the Pyramids on their camels, but many people may not realise they are scores of them, and that amounts to quite a few places where such hidden knowledge, we hypothesise, might be secreted, but in the Coptic Texts quite a specific reference is made to the burying of knowledge by the ancient priests.

The pyramids have been the subject of much human fascination and study and conjecture for quite some time, but as yet, they still have not given up all their secrets. The Arabs, after the conquest of Egypt, were notorious tomb robbers and assailed the Khufu Pyramid with great force and when finding the king's chamber it was disappointingly empty, with no evidence of his ever being buried there.

The very design of them is said to be for a specific purpose and that harnesses and directs telluric or 'Earth' forces. There have even been claims that scale models of them made from red Plexiglas (and one time on general sale), could sharpen blunt razor blades and preserve the cellular tissue of meat. Was it this that kept the mummies in such good condition in certain cases, as well as the preservation process itself?

The amazing treasure found by Carter and Lord Caernarvon was certainly not 'in' a pyramid. They certainly seem to exude an air of mystery and undiscovered knowledge, and it is still not for absolute certain that they were built primarily as kingly burial chambers, in fact it seems most unlikely.

It is said that there are more pharaohs mentioned in written historical works than there are pyramids to house them, which seems also to be a negating factor for that being their sole purpose, but they are surely built on too grand a scale for just the purpose of secreting knowledge, even if they are ideal for such a hiding place.

The 'key' seems to be in the very dimensions, accuracy and alignment of the structures themselves, and their possible cosmic connections with the star system Orien even following the pattern of that constellation in their layout.

With regard to the edifices at Giza, (constructed when Pharaoh Khufu ruled in the 4th Dynasty) and with the advent of the agreed and established 'pyramid inch', used in measuring their accuracy and dimensions (it even allowed for the limestone cladding being still in place that was stripped off in the building of Cairo), a lot of startling facts (or possibly coincidences) occurred, such as adding up the sum of the base sides and arriving at the same figure as the days in the year.

Then there was the height of the pyramid multiplied by a certain number, giving the distance to the Sun, and other seemingly cosmic related data. In fact, the entire structure being seen as a huge data bank or repository of knowledge, and the incredible accuracy of the construction was also confirmed during such research, they seem to be a purposefully constructed 'message' for latter day humans to decipher and resolve.

We usually assign such things as geometrical formulae and theory to the ancient Greeks, such as Euclid and Pythagoras, but then we remember that the Greeks and their culture came after the Egyptians, as the Greeks, under Alexander the Great, were only one of many to conquer the area and, of course, many Greek scholars used The Great Alexandrian Library and studied there and naturally were influenced greatly by the fascinating Egyptian culture. No doubt the rather advanced astronomical assumptions, such as the great gas clouds and 'unresolved stars' they spoke of in the Milky Way were some of the results of that influence.

Perhaps, also, their knowledge of the spherical shape of the Earth and their own seemingly quite successful attempts at estimating its size, and so forth, was attained there. They even seemed to be aware of the atomic structure of matter, no doubt also gleaned from the wisdom of the scrolls from the great library.

As said, Egypt has been overrun by many lands and one wonders at the appalling ignorance, largely through fear of those who destroyed so wantonly, the huge amount of priceless one-off documents housed therein without even so much as a glance at them. Of course, this assumption may be wrong and it may have been the very fact of looking at them that taxed their minds so much they consigned them to the flames all the more quickly. Surely, there must have been some curiosity, a quick perusal to break the monotony of their vandalism? There where learned people existing in those times, yet there was no restraint.

Even the simple acts we all go through from time to time that should take an hour or so, can develop into a full day when we start to have a clear-out of a loft of a spare room. We open and reread everything in our curiosity and what starts out as a clear out becomes a trip down memory lane. Perhaps those chosen to carry out the work were among the most illiterate of the conquering race and could not read anyway, but surely the opposite should have been the case, in which case the astounding facility itself and its many laboratories of medicine and learning should have overawed them.

One would think that to lead a nation of people that person would have greater mental faculties than good rhetoric and leadership, and that even if they did not understand them, others might. If they were commissioned to do so.

In Alexandria in 636 AD, the Caliph 'Omar' destroyed millions of scrolls to heat the city baths, and it is said that it took six months to destroy them all. And a thousand years later, the same thing was happening, this time in South America during the time of the Conquistador reign of terror in that region, as they pillaged for gold and destroyed ancient cultures. Irreplaceable Mayan history was wantonly destroyed by a certain Bishop Diego de Landa. Later on, this person, in realising his stupidity, began to study the Mayan history and seemed to regret his foolish acts.

Why is there this strange proclivity in human beings to fear knowledge and destroy things not immediately understood? It comes down through the centuries to our own time where things not understood become a threat.

Who has not seen the film documentaries of huge piles of books being consumed in the flames of Nazi Germany, or Mao Tse Tung's China? The same adverse reaction to things not understood, is no doubt responsible for the many attacks on assumed UFOs, with no attempts at verbal communication.

If the Egyptian pyramids were conveniently used as repositories for hidden knowledge, much of it would have escaped the hordes of barbarians, who had not yet descended on their lands and since the priests appear to have been entrusted to such a task, they would have chosen the place very carefully.

The Copts of the Middle Ages were quite fascinated by the Egyptian culture and preserved its customs and ceremonies, and their writings of those times speak of a King Saurid of early Egypt also ordering the priests of the day to bury the sum total of all their knowledge and wisdom, including the arts and sciences within the pyramids.

It would be interesting to study and analyse the reports (from those who correlate such things) of the unidentified aerial phenomena taking place or seemingly showing an interest in the area of the pyramidal edifices, assuming a mechanism exists in the region for such things to be reported to. Have they an equivalent to Ground Source Water Watch or Bufora? Mufon?

One wonders if the ET entities, many suppose are present in our airspace, have ever gained access to the pyramids using sophisticated detection methods, in search of the answers we seek. Some abduction claims appear to indicate that the abductors have the ability to pass through solid matter, when appearing in their bedroom to remove them and take them back to their craft, somehow dissimilating the mass of the human bodily atomic structure also, when taking them for analysis.

There is an account of an attempt to carry out a kind of X-ray detection process on the Chephren Pyramid, (constructed some four and a half thousand years ago, between 2780 and 2280 BC at Giza near Cairo). The object of the exercise was to search for any hidden chambers.

It seems that one time Nobel prize winner of Physics, Dr Luis Alvarez, when Director of the University of California's Lawrence Radiation Laboratory, developed a method to be used on the Chephren Pyramid (the second largest edifice) by utilising cosmic rays or muons.[*] The process utilised the sub-atomic particles arriving from space that pass naturally through everything including the pyramids, a cosmic ray recorder was installed in the pyramid at the base of the structure.

The theory was that if the rays passed through any hollow areas or chambers, they would show up as darker areas on the computer pictures. During one period, the recorder was worked for twenty four hours a day over a full year, but it seemed some kind of force, or mystery power, in the pyramid was somehow scrambling the magnetic tapes, and yet when checked, the recorder was said to function perfectly.

The team were quite astounded and said it was scientifically impossible and some force within the pyramid was defying the laws of science. Strangely enough, Egyptian legends refer to Chephren as the 'Guardian' of the fourth Dynasty tombs.

Reference to these alleged hidden chambers within the pyramids were also made by the clairvoyant Edgar Cayce, who could go into a trance and speak

---

[*] Mu-Meson; heavy electron.

lucidly of the mythical Atlantis as though he had actually existed there in some previous existence.

He clearly linked the onset of the Egyptian culture and similar civilisations in South America, to the Atlantean Culture and that its history and achievements, which may astound us in the future, are all carefully concealed within the pyramids.

To be sure, the Egyptian civilisation is something of a mystery regarding its roots and beginnings, and one could use the analogy of the start of the Cambrian Period in Earth's history of some 600 million years ago, when marine creatures of all types suddenly appear with no fossil history preceding them, yet they had skeletal structures requiring much time to evolve, can natural selection explain that?

The Egyptian culture appears more like the end of a lost pre-historic culture than the start of a new one, or the attempt by the survivors of a previous culture, lost in some catastrophe, struggling to revive it again, and little historical, natural chronology, is evident of the culture arising through a long time period as should be evident.

When the pyramids stood in all their refined glory, smoothly clad with bright lime stone finish and said to have been capped with gold sheeting, no doubt together with the equally adorned surrounding buildings and edifices, they must have looked like something from another world, to someone beholding them for the first time, considering the dwellings, and how the rest of the world lived at that time, and no doubt the 'locals' with their ripping off the cladding for building materials, could be seen as equally barbaric as the destruction of the aforementioned parchments and scrolls, clearly they had no regard for the beauty and wonderment they surely must have evoked, it was desecration of their own culture.

How would the world react if the modern day Greeks pulled down the Parthenon to build houses? One wonders if they had had the skill at the time so to do, whether they would have pulled the actual pyramids themselves down for the use of the blocks.

If the mythical Atlantean Culture did exist, then all history and trace of it disappeared in the alleged catastrophe mentioned by the Egyptian priests. One wonders how that culture could have arisen. Was there a cosmic connection? If alien intelligence has been observing Earth as long as the signs seem to indicate (well into history) then they will know more about human history than anyone on Earth will ever know.

One thing is for certain and that is the fact that any culture, no matter how advanced, unless they could work miracles, would almost certainly degenerate very shortly after being deprived of its cultural advantages, particularly in a harsh environment with perils and predators all attempting to eliminate it, not to mention geological and seismic activity. They would also see their numbers reduced by unfamiliar bacteriological strains. But if the intelligence or brain development was already achieved or highly evolved, no matter what degeneration took place and for how long, the inherited memories of past glory and achievements would gradually emerge and come to the fore, and no matter how tribal, savage or hostile the survivors who proliferated eventually became this mental striving for greater achievements and advancement would continue until another marauding tribe or a geological catastrophe came along to annihilate it, and this is precisely how Earth's history and its peoples have come through their long struggle to achievement, followed by degeneration through the centuries, human behaviour rather than catastrophe being the cause. Every indication seems to point to the ever continual struggle of the neurological process of positive and negative forces in a state of equilibrium in the brain itself; in mentally stable people the forces remain in balance; in the criminally insane these forces go wildly out of balance. We cannot switch on our television sets without seeing a reporter nervously ducking as he tries to relate some battle or carnage going on somewhere on the globe, as humans proceed systematically to destroy other humans. Yet we quietly lay plans for star-ships and voyages to other worlds, with plans to 'condition' them for our life form. With regard to the alleged alien presence, 'they' may have finished those studies long ago, and know exactly what to do. It would appear that the current abductions are centred around invasive operations of a gynaecological nature. Are they producing, from removed female eggs, some kind of 'quasi-human' entities equipped with advanced alien neurones, but human in other respects, to be purposely inserted (or may be long previously inserted) into all the important and persuasions in human life? Their ultimate purpose may be to offer criminals and prisoners neurological 'enhancement' to eliminate all their negative behavioural traits. For this they would be freed, our own medical scientists and geneticists will be carrying out such activity in the future if left alone to do so, particularly if there is a lessening of influential ethical restraints, but no matter how beneficial an option may be, there is always a group that will oppose it, but if the majority do support it, then let it happen that is democracy.

Anyone skilled in the knowledge of our comparatively simple transistor radios could pick one up from a tip, casually discarded by someone without the skill, and was not too bothered anyway as they had replaced it with a desired more modern type. The radio expert, if he was so inclined, could probe into the discarded item, and with a bit of bridging and perhaps a small component replacement, could no doubt bring that unit 'back to life' so to speak. We could compare a mass of electronic circuitry to a basic brain. Even in the Bible Jesus said "What I can do ye also can do". I said in *Pillars of Fire* that cases are on record of special powers suddenly becoming apparent in people who have fallen and received a blow to the head. This implies that the necessary circuits were 'bridged' in the actual impact that caused the activation of the neurons, perhaps latent and available in all of us that caused this ability to come into play. Our medical scientist of the future may well be able, quite easily, to cure all congenital mental disorders, simply by locating the correct area of the brain and carrying out certain 'adjustments', or what we call 're-wiring'.   Mental aberrations seem more prevalent and widespread today, perhaps due to the pressures of modern life, and little mentioned in accounts of ancient medicine. The ancient Greeks were very proficient in medical cures and possibly gained much from the Egyptian knowledge.   Solon learned much from many discussions with priests of Sais. He also discussed past worldly catastrophes and the legend of Atlantis with them. But it was sufficient for the Egyptian priests to outline the loss of the landmass and speak of other Earthly disasters, but whether the achievements of the probable forerunners of the Egyptian culture were told to him (or even known by the priests) is debateable, in any case they would not have wished to hand over all their knowledge to the Greeks on a plate, they considered that the Greeks where still in the learning mode, so let them learn.

Although certain priests were instructed to bury the sum total of their knowledge, there may have been other data inserted into the containers that even the priests were denied access. Gold seemed to be plentiful and well used and such a noble metal would almost certainly have been utilised as the containers for such knowledge, in order for it to last as long as possible into the future, as they knew the pyramids would do so.

The clairvoyant Edgar Cayce's descriptions differ somewhat from Plato's account, which is said by him to have been records kept by the priests of Sais and painted on the temple columns.   These were shown to Solon, Plato's forerunner, and seemed to have been for general consumption, but it would appear that Plato did heavily embellish them, and it is suggested that his inspiration in writing of Atlantean harbours and constructions, and the commerce thereof was influenced by his knowledge of such in Tyre or Phoenician and Cretian cities and ports, and possibly the ill fated island of Santorini.   (THERA), but this latter assumption suggest Plato would have stayed within the Pillars of Hercules not beyond them.

After the excavations in Crete, buried by the explosive volcanic outburst of the Island of Thera (or Santorini) long ago, the finely preserved frescoes murals and artefacts seem to bear this out.

Aristotle, regarding Plato's account said, "He who invented it, killed it". However, other convincing geological, and some written references seem, to strongly reinforce the existence of the island. As said, the Conquistadors were told by the Aztecs that their ancestors originated on Aztlán, a submerged island to the east.

Another of the many similarities one can point to regarding the coast on both sides of the Atlantic, are some similar words, for example, the Berbers of the African Atlas Mountain region (who are not Arabic) use the word 'Atl' for water, as did the Aztecs.

Beach sand only forms by constant and relentless erosion by the sea (at sea level), and when one finds beach sand in oceanic samples, as is the case near the Azores, this is seen as rather compelling evidence for submergence of a land mass just as the beaches at high altitude around the Lake Titicaca suggest the heaving up of the coastal region impinged on heavily by the 'geo/tectonic' Nazca Plate, and once again this geologically turbulent happening of some 12,000 years ago arises, which may have greatly influenced such upheaval and the associated geological disturbance.

It is often suggested that the Canaries and the Azores are the remnants of the mythical island and the Guanches, the ancestors of the Canary islanders, are reputed to have related their traditions of the disaster that befell it to visiting Spanish seafarers, and that their predecessors where Atlantean survivors.

Much under ocean movement still takes place in the Atlantic and cable laying and repair ships have noted the depths of their cables alternating considerably, even breaking.

There are some strange habits of the migratory birds during their crossing of the Atlantic, that stop and fly around in great circles over the supposed area the mythical island occupied, then go on their way as though some racial memory they possess told them something. They also do it on the return trip.

Great shoals of eels all swim west into the Atlantic and submerge at the Sargasso Sea, a great seaweed-filled mass, said by the Atlantologists to be the seaweed, grass and vegetation, that floated up after the catastrophic sinking of the mythical land mass, and has proliferated ever since.

It is stated in seafaring tales, that many a ship was trapped there until the crew expired and the hulks disappeared into the depths.

The Atlantic has had more than its share of sightings of unidentified craft entering and leaving the sea, and large dull grey circular objects have been sighted on this Ocean. It would appear that a crew member of the original

Queen Elizabeth liner saw a 'plane' of some kind flying directly toward his ship, then suddenly disappeared into the sea, a hundred yards or so from his ship, with no splash, oil, noise or wreckage.

A silvery object was seen to dip into the sea off Daytona Beach Florida, in 1935 with red and green lights. Two witnesses, Mr & Mrs Forrest Addington observed it from the Hotel Daytona Beach, as they watched the sea from their balcony. Coast guards searched, airports were checked, no missing planes were reported.

However, it has to be said that to have all this technology located in one precarious spot that is over the mid Atlantic Ridge did not imply great knowledge of seismic or volcanic forces and their destructive capabilities. Having said that, of course, people still carry on living in areas known or threatened volcanic or underground fault activity, such as San Francisco and California in the USA and adjacent to many volcanoes. An old gentleman made it clear that nothing would get him to leave his home on the mountainside of Mount St Helena and died when it blew apart.

If such a cataclysmic event did occur, one wonders just how far the survivors did travel. As well as the Nile Delta and the areas of South America, perhaps the ancient Indian legends that speaks of 'Vimanas' (flying craft) and seemingly nuclear warfare in the Mahabharata also had their share of 'Atlantean' settlers wishing to get as far away from the sea as possible, even as far as the Tibetan Range.

A good deal of ancient writings, which appeared in the past to be imaginative tales, do not seem with the advent of their technology quite attributable to this category. Indeed, some seem frighteningly realistic in terms of advanced weaponry techniques. Many things around the world astound science and cannot be adequately explained, but because they do not fit neatly into our prepared niches and expected chronological order, they are quietly ignored. One would expect it to be the other way around and a challenge to our science to do our utmost to explain them, but the human element prevails, no historian would wish to dump all the pre-conceived and written data.

The continual and persistent reports, alleged encounters and abductions, with regard to aerial phenomena are still with us. there are many fantastic edifices around the world that defy all explanation of their construction, not only with the lack of technology, we assumed was the case in those times, but even with today's method and ability.

The trend today is to assign an astronomical reason to most of them in order to bring them within the realm of human activity in those times with regard to the seasons, growing and planting and survival of the ancients. Then it all seems logical and no more awkward questions should be asked.

Perhaps it was unfair to generalise, with regard to science ignoring things and hoping they will go away, certainly when a scientist such as Maria Reiches, who devoted an entire lifetime to the study of the Nazca Plain markings in Peru. She was trying her upmost to explain the unexplainable, though it is hard to imagine monkeys, birds and spiders as having any astronomical connotations, but there are other designs and shapes (all of which can only be properly appreciated from an aerial viewpoint), which could be seen as such. The many and varied lines that go on for huge distances, up and down over obstacles, but still straight, would be bound to point to some star in the night sky, no matter which direction they pointed, and even if they did not, one can always fall back on the 'Precision of the Equinoxes', meaning that they would, no doubt, have pointed towards some constellation, even if they do not today.

Certainly, there are many writers that have visited, evaluated and gazed in wonder at such enigmas as the edifices at Tiahuanaco in the Andes. It is difficult enough to breathe there at 13,000 feet let alone do heavily physical labour and construction work, including farming and agricultural labour in such hostile conditions, yet terraces indicating that kind of activity do exist there, but so does what appears to be a port and a 'salt line' and other indications of the area once being at sea level, around the area of Lake Titicaca.

If our science cannot explain things like this that puzzle us, it is unwise to deride theories that suggest what may have happened so long as they are scientifically feasible. The stones that complete the edifices are not just laid one on top of the other, but have many curves and angles and seem to confirm loosely to the analogy of wet sandbags placed to make a wall where the weight and malleability forms in patterns. It is even suggested in some legends that the ancients knew of chemical methods and substances to soften stone.

Those huge stones could not possibly have been put on and off a 'trial and error' fashion as stonemasons chipped and filed here and there. They fit precisely with no gaps and weigh many tons each. How could they have been transported in such rugged terrain from their quarries, hundreds of miles away over mountains, gullies and rivers? It would have been difficult enough on a flat surface at sea level. Ancient legends (those regarding Merlin for example) suggest the ability to 'levitate' large masses and make them 'fly through the air'.

There are indications of the work being suddenly abandoned, just as there is on Easter Island, with those huge statues of 'most unpolynesian' beings, erected, fallen or half finished here and there.

Chapter 2

WHERE WOULD "THEY" COME FROM?

Theorising on there being some 95 million sun-like stars in our galaxy is of little use to us when considering who may have reacted 'out there' to our mass communications, even though the radiations may be charging along at an enormous speed, the sobering thought, is the galaxy is 100,000 light years across and we are some 27,000 light years from the centre. We would have to look much closer to home.

As it happens, there are seven 'sun-like' stars within a distance of two dozen light years away and, by now, if any intelligences do occupy any of the supposed planetary systems, perhaps one of them has the technology to be able to detect our give away signals. If so, they could have been analysing them since they first reached out to their region of the galaxy. It is fair to assume that at least one of the possible retinue of planets circling those stars has the necessary prerequisites for life to have evolved, but perhaps not an intelligent creative form we have assigned to our visiting ETs formulated from witness's accounts and those of alleged abductees.

The names of these comparatively nearby suns are:- Epsilon Eridani a 'mere' 10.8 light years from Earth, Tau Ceti, 12,2 light years away and Sigma Draconis, 18.2 light years away from Sun. These three stars have some 80% of the Sun's mass.

Delta Pavonis is just under 20 light years away and 82 Eridani is close to this distance from us. They are very sun-like, with over 90% of the Sun's mass. Beta Hyri is 21.3 light years away and is some 20% above the Sun in mass, and Earth's radiations have been streaming over any worlds that may exist in those regions for quite some time.

If we take Epsilon Eridani, for example, any intelligence with the ability to receive, would have knowledge of all the mayhem and conflagration occurring on Earth from the Second World War to the Gulf War, which would include the Korean War, Vietnam, the breakup of Yugoslavia and numerous other conflicts and, by analysis of our quality TV broadcasts, would also have seen the positive side of our nature. However, by studying certain productions on history and programmes and features such as, *The Great Commanders*, they would be quite aware of our proclivity toward conflict and be able to observe that we had been at it rather a long time. Of course, to extract the faint whisper of our TV programmes, they would be required to have extremely advanced and sophisticated detection equipment. Perhaps an alien's summary of our nature in this regard may be: - "The greater part of the economic effort and expenditure since the mining of metallic ores in that world has been toward the manufacture of more efficient weapons with which to slaughter each other". A rather damning statement, but nevertheless true of human endeavour. If, as previously suggested, they perfected their receiving apparatus and controlled their emissions, they may be congratulating themselves on their wise decision to blanket their outgoing radiations and concentrate on listening out. Did we have the chance of computing where alleged aliens originated? In the famous human abduction case of Betty and Barney Hill, which will be related in Chapter three, Betty Hill recalled asking alien entities various questions and, at no time did either of them actually hear spoken words or even see their lips move, although they did receive mental communications.

They do not recall seeing a mouth at all, perhaps just a thin 'slit', and this description, together with the mental communications, seems to be one common factor in all abduction cases. As previously, stated, human beings may themselves one day fit this description at some point in the future (should we have one). To turn to the convincing case of an abduction related by Betty and Barnie Hill.....

In questions to the alien entity, who seemed to be the leader, Mrs Hill asked from whence they came. The reaction to this was the displaying of a star map and the communication from the alien, "Do you know where your Star is on this display?", to which Betty Hill replied in the negative. "Then it is little use of showing you where we come from", came the reply. Now, where Mrs Hill had no idea how to draw or reproduce this star display in a normally conscious state, under hypnotic regression she could and, interestingly, after later study by a different party, the star layout did seem to match roughly to a group seen from Earth, but one supposes that with so many stars visible in the night sky, anyone's drawing of a group of stars would roughly match some cluster in existence.

However, the point I would like to highlight is, are we only concentrating on star groups as seen from an Earthly viewpoint? Is it possible for astronomers, with their sophisticated computing techniques, to compute what a star group with Earth's Sun in it would look like from Tau Ceti or Epsilon Eridani or any number of fairly nearby star systems where the alleged 'abductors' could have originated? If so, we may find an immediate and obvious match with the star map extracted from Mrs Hill, thereby indicating their system of origin.

Although, as said, we have so obviously announced our presence in the galaxy to any listening agency with our outpourings, it is a very large assumption to expect alien intelligence sitting there twiddling their tuners and aligning their dishes in the sense that we do with our radio telescope assemblies. They may do it, or achieve the same end in a way incomprehensible to us.

To expect them to do things like we do or to have developed as we have, would require so many coincidences, so many similarities in biological and natural development as to make it seem very unlikely, but with regard to the humanoid configuration evolutionary and natural selection process may strive to produce it as the ideal carrying medium for an advanced intellectual capacity. It may logically follow then that, if there are Earth-like worlds and bipedal humanoid entities have evolved with intellect, a lot of things would possibly occur as here on Earth if they had imagination, reasoning power and the ability of abstract thought and creativity, which would all naturally be expected with such intellect, then the intriguing sight of a rainbow, for instance, should be the starting point for them, just as it was for us toward our first radio telescope via the breakdown, analysis and study of light. This assumes that their planet has the necessary prerequisites to produce a rainbow, i.e. an Earth-like atmosphere, precipitation and so forth.

Of course, there could be planets circling sun-like stars that equate almost exactly to Earth, yet having entities nothing at all like human beings residing thereon. We only have to consider that the Earth was totally devoid of humans and was ruled entirely by the dinosaurs for a period of close on 200 million years.

One wonders how many (if any) alleged UFO or aerial phenomena visited Earth's skies in those times. to be sure, there would have been very few landings, but if an alien race, desperate for a new home, saw such an attractive planet, populated by such creatures, they may have been very tempted to rid the world of them to pave the way for their own occupation.

Some anthropologists have suggested that the so-called 'branching off' from the simian kind that allegedly occurred in regard to the evolutionary theories, may have occurred as long ago as 50 million years, in other words, close to the origins of the true Pongid apes. Now when we consider that the dinosaurs did not all perish exactly 65 million years ago but that some lingered on for quite a time afterwards, we find that the appearance of the human entity could have occurred as a natural follow-on from the dinosaurs. This is not to suggest that humans in any way evolved from them, but it seems to suggest that their demise was a prerequisite to the appearance of the human entity. Having said that, an anthropological museum in Canada produced a reptilian bipedal entity with other humanoid features that could have evolved from the dinosaur era.

It is perfectly possible that the human body-form could have its origins elsewhere in space and that we are all descendants of a race of cosmic refugees that arrived here, coincidentally, shortly after the dinosaurs' demise, or directly contributed to it, prior to landing here. This would of course explain the advanced human brain that in accordance with current evolutionary theory does not seem to have had enough time to develop.

There are legends from the people of India's North East Frontier that relate to space beings arriving on Earth in time long past. They even mention the area the Long Kapur Hill in the Lohit Valley.

There is another Earthly legend from an island in the South Seas in the Tuamoto Group that sounds exactly like an account of a long space journey coming to an end with an arrival on Earth. It says, "In the beginning there was only empty space, neither darkness or light, neither land nor sea, neither sun nor sky. Everything was a big, silent void, then the void began to move and turned into Po (Earth). Po itself began to revolve (orbit), new strange forces were at work (gravity), the night was transformed, the new matter was like sand, the sand became like firm ground that grew upwards (the landing). The Earth Mother revealed herself and spread abroad and became a great country".

It could not sound more like a handed down account of their ancestors' arrival in a spaceship to a new world, i.e. Earth and it just may be possible that the early evolution of the human race occurred on a world very far from Earth, and our early bones will never be found here as they may reside in the soil of another world that may now be a roasted cinder circling the remnants of an exploded star.

It would appear then, that there is a very wide area of choice from which to assume our hypothetical alien entities have come, particularly with over 5% of the stars in the galaxy having roughly the same mass and temperature as our Sun. This gives an enormous amount of assumed planets and it is reasonable to assume that at least one planet of each star has life-giving conditions. Nevertheless, it is quite surprising how the odds whittle down when we start to consider how 'just right' everything has to be for multi-cellular life forms, such as ours, to be able to emerge.

The most abundant stars in our galaxy are 'Red Dwarfs', which are, in fact, ten times as frequent as our sun type stars. Many stars burn themselves out far too rapidly for life to emerge on their assumed planets, but that, of course, alludes to 'indigenous' life there, and it is perfectly possible that intelligent life could have arrived on any world that has at least evolved a primitive atmosphere. They may have even manufactured one for themselves. Certainly, a life-supporting world would not be passed by simply because its star was long into its main sequence by entities looking for a planetary home. A stars' lifetime depends wholly on its mass and the heaviest burn the brightest and live the shortest lives. To be sure, if the Sun's mass were just 25% more than it is, it would have already burned out.

It is really not so farfetched to suggest that 'they', being hypothetical aliens arriving on a world with the basic prerequisite for life present, to be able to 'molecular engineer' any situation they wished.   Certain scenarios are envisaged, dealing with the advances and discoveries, in analysis of the smallest particles, and the research, in such places as Silicon Valley in the USA, specify that at the present rate of 'time versus miniaturisation advancement' shown on a graph, eventually we will be pushing the very atoms themselves about into any form we choose, changing or converting matter by altering its nuclear assembly, that is, its atomic number.

The most advanced microscopic analysis sees the fuzzy motion of atoms and, when freezing the item beforehand to slow the motion, the actual individual atoms have been attracted or jumped on to the pointer. This seems to apply a process where one could alter the 'atomic number' of materials and build anything we chose. We would become virtual 'creators' on Earth. To be postulating on this type of activity when certain human beings on certain parts of Earth, such as the Amazon and New Guinea are not uniformly advanced, and this situation could exist on other worlds.

We may postulate on whether the intentions of the possible alien entities in our airspace are strictly honourable, but how honourable would ours be, if at some future time, when our technology had advanced to the point of our travelling to other star systems? If we had seriously plundered the Earth's resources, then found an Earth-like world devoid of seriously predatorial creatures, we may be tempted to take it for ourselves, regardless of the occupants.   This imaginary world may have an abundance of mineralogical resources for the taking, even if it did have seriously savage creatures who would attack us, the commander would say, "It's them or us, I vote us anyone opposed?" . Just as we assured ourselves in our write-up, written by men for mankind (yet assigned neatly to the Almighty), we conferred the 'Dominion over all other life forms' on ourselves, we may just decide to extend this dominion to those other world creatures.

Most of the obviously fierce creatures of the Earth have been hunted practically to extinction, or placed in our zoos, but there are other creatures of this world who could very easily terminate our dominion over them quite easily, by terminating us without any help from hypothetical alien entities, if they seriously increased in numbers instead of running away from predatorial humans turned on them and hunted them down.

The distressing sight of oil-soaked sea birds struggling pathetically out of polluted waters seems to be appearing on our newsreels with monotonous regularity. In some countries, birds are snared, shot at, captured and even tinned as edible delicacies. We shoot many varieties when the season begins and eat many of the other bird species, ducks, pheasants, partridges, etc. The birds may just decide that they have had enough of human behaviour toward their species and see us as a direct threat to their survival and attack us in their thousands in some 'Hitchcock' type of nightmare scenario. What of the unseen and largely ignored problem of the rats. They take such pains to keep out of our way and lurk unseen in the lower regions that clearly they regard us as a threat. With their numbers, seriously proliferating, sufficient sources of food must be starting to become a problem for them. The surface of the planet teems with food. Should they ever decide to take it we would not stand a chance. There are probably a lot more rats than there are people. If they were ever panicked or seriously disturbed by, for example, the development of a gas or agent rather like fly spray that is harmless to us but attacks wasps and flies like nerve gas on their metabolism, and if, in some future impetuous, or ill thought out action, we decided to reduce their numbers by pumping such a product in great quantities into the sewers, they may erupt out of every manhole like burst water mains and, in panic, attack and devour everything in their path. As a conservative estimate, there are probably around a hundred rats for every person on Earth.

A recent television programme may have come as a shock to some people where it was shown that nuclear bombs were exploded in space for little other reason than to see what would happen, but cloaked under some 'environmental control' type of respectable title. Test bans in space put a stop to it, but science still beams high-energy fields into the Aurora Borealis in Alaska and ensure, with tight security, that nobody really knows what they are up to. Even when the very first nuclear devices were detonated, the scientists could not guarantee that they would not set the entire atmosphere alight in some kind of chain reaction, but they went ahead anyway. We must say though, that they had little choice knowing that the Nazis' where developing them, so, if anyone is going to destroy the world, let it be the allies.

Even today, some countries, with regard to test ban treaties, refuse to sign them. The Aurora Borealis is a quite natural phenomena of the particles of the solar wind reacting with the Earth's magnetosphere, which protects us from the excess radiation from the Sun. By the time, we get around to caring for, rather than trying to control the environment, it may be too late. We quite often seem on our newsreels, the authorities lifting the protestors bodily out of the way of the tree wrecking bulldozers in our haste to concrete over practically the entire world. Strangely, a lot of us come down on the side of the authorities and view the protestors as the 'the great unwashed', or as troublemakers. Perhaps this is because they are usually Hippie types, or New Age Travellers, many of whom, like Gypsies, seem to feel they can go anywhere or on to any land, they feel like. Even Greenpeace activists and Animal Rights people are viewed in the same way – all looked as somewhat odd (with regard to the latter, it is sometimes self-inflicted when hearing of their more extreme behaviour). However, we should all be thankful that such organisations like 'Greenpeace' do exist and that at least some of us care enough to do something positive about damaging activities to our environment. They bring adverse publicity to adverse behaviour. However, there must come a time when the authorities will have to listen up in order to save the Earth. The greatest polluters such as the USA and China will have to take the lead in this scenario.

Any imagined alien entities in Earth space may possess extremely sophisticated computing equipment, and perhaps when feeding all the data of the more questionable human behaviour with regard to past, present and future actions and plans relative to the environment of Earth, the printouts may not bode well for their decisions regarding us. The question may be, will they stand back and allow us to destroy such a probably extreme rarity in the cosmos as a life-giving planet? Any obvious lack of regard we show for Earth may hasten their plans to relieve us of it, particularly if their own world is under serious threat, perhaps due to no fault of their own. They could occupy our developed and desirable areas and herd us into deserts, and underdeveloped regions, putting us back thousands of years in our development. However, with regard to our own destructive activities, groups like 'extinction rebellion' are bound to exist as our activities will be passed onto them to deal with in their future.

The sages of old who made such comments as, "We do not inherit the world from our ancestors, we borrow it from our children", made them long before we had seriously commenced to attack the planet with modern day pollutants.

Chapter 3

ARE THEY ALREADY HERE?

Who are these strange entities that have been labelled in UFO lore as M.I.B's or Men in Black? Many cases are on record of people who have had their seemingly authentic sightings of the UFO phenomena published, receiving visitors that have an odd appearance that makes them appear as some sort of misfit or crude human duplicate.

The usual description is, they appear as some sort of misfit or crude human copy. The usual description is, they wear ill-fitting clothes, not always fashionable, usually black, speak strangely, usually arrive in large black limousine type vehicles and make some warning or other regarding the dire consequences of relating the experience the UFO observers have had, to anyone else. These threats are rarely carried out, and as far as is known (with one or two exceptions), nobody relating a UFO report has come to serious harm because of ignoring veiled threats. These early M.I.B. were very different from the current 'Hollywood Hero' M.I.B.'s of today. Oddly, Betty and Barney Hill had no such 'visitation'.

Many cases of alleged human abductions are on record. The famous Betty and Barney Hill case, were a mixed race couple encountered a bright moving light while driving through the mountains of New Hampshire on their way home at night, is the most authentic.

Even the doctor who hypnotically regressed them and brought out a fantastic story of abduction and analysis of both of them in separate areas, on a spacecraft, could not seem to bring himself to believe it all, in spite of the dramatic reactions of Barney Hill while under hypnosis.

Their stories had been of an experience so vivid (and each checked out against the other) that they came close to emotional breakdown when relating them in a 'regressed' condition and as they had kept their experience to themselves for so long, it was clear that they had not sought any kind of notoriety.

One interesting feature of Betty Hill's alleged alien analysis was a gynaecological test that only later came into general practice in that branch of the medical profession. In all cases of abduction, human bodily material is removed. In hypnotic regression and polygraph techniques, it is always emphasised that any information gained is only what the 'victims' (or perpetrators in the case of lie tests) believe themselves to be true and, sure enough, the doctor in the case of the Hills' experience believed they had dreamed it all and after much discussion of it between themselves, began to share the dream. He was obviously a rational person and could not accept the 'living aliens' part of their encounter but strict rationality can close ones mind and the old 'can't be so it isn't' factor emerges.

Of course, this could be a classic case of the rationally and objectively thinking analysers, not wanting to accept something so fantastic, but they choose that way of thinking because of their knowledge that the human brain, fantastic an instrument through it may be, can indeed be fooled, as can our visual apparatus or any of our senses. The human brain is a very fragile organ in the abstract mode and can be easily 'manipulated'.

The Hill case seems to be most profound on record and, if we do not believe that one we, cannot believe any of them, however, when the figure reached over a million of alleged worldwide abduction claims the world had to listen.

The seemingly human quality of disbelief of what in some cases seems to be rather convincing circumstantial evidence of alien presence, could be used to great advantage by any alien entities existing in reality, after all if, going back to biblical times, their forebears were responsible for the activity so 'extra terrestrial' in nature, our belief was so different from the reality. This alternative belief was used to great advantage by E.T. masquerading as 'Angels' to achieve their ends.

In the early days of flight, when aircraft started to fly faster and the power of static build-up due to the friction of air passing over the metal surfaces was more obvious, static 'wicks' in the form of paintbrush like 'appendages' protruding from the trailing edges of the wings were necessary to dissipate static electricity. It was necessary to have each component 'bonded' to each other, either by metal to metal contact which was then sealed to prevent ingress of moisture or, by braided wire bolted to each part to ensure a good flow of the static charge out toward the dissipation points. If these bonding wires became detached or broken, the aircraft radio would be badly affected by static background noise. It was also necessary to have electricity conducting compounds incorporated into the aircraft's tyre rubber to ensure final 'earthing' on landing, before this implementation regarding the tyre rubber, people who made contact with the landed craft where blown on their backs as the built up charge 'earthed' through them.

Perhaps the well know 'Foo Fighters' allied aircrews experienced in World War Two were a manifestation of this type of phenomena, dissipating from the craft itself small bright balls of light that may have been a form of ball lightning.

Of course, it is not only in Earth regions where aerial phenomena have been observed. Astronauts have seen them in Earth orbit, and going to the Moon and also in the Lunar environment, and I have related much of this data in *When the Moon Came*.

If at least a few of the older, or even more recent, cases are actual alien craft, what kind of time period or observation time scale can we ascribe to the alleged alien interest in us? Some of the descriptions in biblical writings would convince many (and have done so), that they were real ET phenomena. Over 3,000 years is a very long and patient exercise in human observation and analysis, but then, from what we hear of the strange effects of time dilation (depicted well in a successful film, *Close Encounters of the Third Kind*. Where the lost Flight 19 pilots returned younger than their offspring), perhaps the time period we consider lengthy regarding their observations, is just a few missions to them. The feeling remains as strong as it ever did that somehow 'they' are connected, or at least their forebears, had some profound connection with human beginnings even if we do insist on assuming primate lineage.

Even if the apes existed for another fifty million years it would appear to be extremely unlikely that their brains would have developed into anything like that of humans, unless some 'exterior entity' with the necessary genetic skills and knowledge, could circumvent such seemingly required time periods and the indication of some form of intervention or manipulation by some 'unknown' variable definitely exists with regard to our own development.

The human body form seems then, to be the ideal arrangement to match the advanced and highly evolved brain and that this may have been the aim of any alleged ET 'biogenetic' process. Spiders are ideally constructed for their tasks, sharks, dolphins and whales, moles in the earth – the list is enormous – all suitably equipped for their environment, moulded by evolutionary forces to ensure their survival.

Even though apes cannot stand upright for long due to their hip joints, what vague humanoid form they do possess seems wasted on them and would be even more wasted if they achieved fully erect stance, without the necessary neurological endowment, their own brain would be insufficient to provide the impetus for them to fully utilise their construction in a creative manner and 'homo erectus' would be a first-class example of this.

If humans did evolve from apes as a natural process but only remained semi-bipedal, there is no good reason, given our same over-endowed brain, that we could not have arranged our advancement and technology to suit our body form with regard to our leisure and sporting activity. Although cricket would be out, we could manage rugby quite well, and even if possessing only partial bipedalism, this would not retard human creativity.

No one on Earth today can do other than theorise how the enormously over-endowed and developed brain could have come about in such a comparatively short space of time, and those who theorise the possibility of ETs short-circuiting plodding evolutionary processes have an opinion as valid as any other. Nevertheless, statements such as 'biologically impossible' are made and, of course, this is absolutely true for humans for now, but when we consider the long list of things once viewed impossible that now are entirely possible, it indicates the folly of such pronouncements.

However, what if the enormous strides in genetic science we are witnessing today had taken place thousands of years ago in any hypothetical ET domain? In spite of these categorical denials of such possible ET activity, those making them do (graciously) admit the possibility of more advanced worlds, with beings having this ability, could well exist.

One could not imagine groups of reporters continually loitering outside remote radio astronomy facilities on the premise that a number of the team would suddenly pop his head out and say "OK, this is it boys, the scoop you have been waiting for. Press conference in half an hour". They just quietly get on with their work in their often-remote regions and any such profound signal coming in would probably see the chief scientist reaching for a 'red' phone, rather than dialling the editor of some national newspaper, and they may well have it made clear to them that they must, if wishing to exist financially.

Some of the explanations of aerial phenomena sound a bit trite simply because they are used to often, i.e. swamp gas, ball lightning, the Planet Venus, meteorites and so forth, but this would be the case as such natural phenomena are manifested in our skies so often, especially the release of weather balloons but all these so-called explanations are inadequate in many cases.

Anyone seeing a certain type of meteor immediately knows what it is. That is the type that make a brief bright trail through the night sky in a second or so, appearing as a streak of white light that quickly dissipates. Meteors do not always approach like a car coming towards us on the motorway. They sometimes come in like a vehicle overtaking from behind and, if they approach just a little faster than the Earth's speed of orbit and come into the atmosphere with the direction of Earth's rotation instead of against it, slow yellow fireballs rather than a white hot streak, could manifest themselves, and be quite prolonged and such photographic evidence of these does exist and probably gives us all the meteorites we possess.

The various orbiting satellites are often quoted as mistaken for UFOs. Of course, such bodies reflect the sunlight and appear very much like a moving star and hold a steady course and, obviously, could not be offered as an explanation for an object that was seen to alter course or even make 90-degree turns and change velocity abruptly.

One would expect though, that clever alien craft would take steps to prevent such reflections of their craft, the opposite seems to be true and may be all part of the proclamation in the Bible that before the second coming "There will be signs in the heavens".

The astronauts, not surprisingly, with all their alleged encounters, particularly during the Apollo missions, seem singularly united in their belief in cosmic entities. Frank Borman is convinced we will encounter extra-terrestrial life in the future. Astronaut McDivitt said, "They're there, without a doubt, but what they are is anyone's guess".

Then there is the case of Mariner 4 picture signals from Mars being supposedly interrupted by a UFO. An Australian tracking station at Tidbinbilla noticed the signals became unexpectedly jammed, and about 30 miles away a UFO was hovering and giving off a steady bright light. When Australian Air Force jets went up to investigate it the UFO disappeared and the Mariner 4 signals recommenced.

Many quotes have been made by eminent people about possible extra terrestrial presence in our skies and suggest that the real possibility of their being extra-terrestrial nature does exist. The US Air Force Academy of Colorado published the following in the Volume Two of 'Introductory Space Science' textbook:- "The most stimulating theory for us is that UFOs are material objects which are either manned or remote controlled by beings who are alien to this planet. It is suggested that at least three, or maybe four, different groups of alien intelligence could be visiting Earth. We should not deny the possibility of alien control of UFOs on the basis of preconceived notions". This is quite a contradiction when to have appeared on TV and newsreels totally denying anything that may suggest they are extra terrestrial. Some people have suggested that this anomaly is all part of a contrived plan under the 'black propaganda' and fake news, released that are so prevalent today and whole plan is to sow the seeds of doubt and confusion and making people give up worrying about it. This then takes the pressure off the Air Force to have to continually trot out denials and explanations.

Dr W Howard, late of Harvard University stated, "Intelligent beings abound in the universe and most of them are far older than we are". One wonders what kind of information has he had access to, in order to be so positive in his statement.

A quote from Javiar Garzon, who was a physicist on the staff of The National Astronomical Observatory of Mexico City states:- "The UFOs really exist and apparently come from other planets" these statements are quite positive in their nature.

Many support the previously stated possibility of 'genetically produced' entities living on Earth. They would be extremely hard to detect and may have some human pedigree and be born of human females with memory implants to do the bidding of their masters whether they were aware of their alien background or not. They would automatically carry out their 'instructions', whether they knew or not that they were hybrids, technological progress and discoveries may have been 'guided' by such entities through successive generations. Their longevity is indisputable and recorded throughout history. Which may be significant and certainly most people are aware that sightings of unexplained phenomena abound through history. In England in 1646 there were reports of 'ships in the ayre' recorded in a book in the British Museum Library regarding aerial phenomena seen in Norfolk and Cambridge ... "Betwixt Newmarket and Thetford", there was observed, "a pillar or a cloud to ascend from the Earth with the bright hilts of a sword towards the bottom of it. The pillar did ascend in a pyramidal form and fashioned itself into the form of a spire or broach steeple" .... It would appear the witness could no more describe what he had seen anymore than Ezekiel could in biblical times, but the old reports all sound like cosmic modern day activity to us.

However, if they were manifestations of alien activity, the occupants or controllers found Earth easily enough then without the help of any radio emissions, as clearly, in those times, we were not emitting any.

If and when we do finally establish a lunar base and do some useful mining there, we could utilise helium 3, mined out from the lunar landscape. But we seem to be going ahead of ourselves and talking more of Martian colonies and terraforming that world, when analysis of the Moon and its returned minerals seems to indicate that it is there that we will find the substance to power our spaceships to the stars.

Fusion engines would, over a lengthy period, power up an interstellar craft to enormous velocity. To be sure, enormous velocities will be needed if we are to have any hope of even reaching Proxima Centuri, our nearest cosmic neighbour, but even at a ¼ of the speed of light we could get there in 16 years and it is a safe bet volunteers would be found. What a sight to behold standing on a planet there with three suns in the sky.

Imagine a place you know of near your town that is five miles away, then say two words, "one thousand", and if you have reached it in that time you would have travelled at 18,000 mph, something around the rotational speed of the Earth and a space vehicle before its 'burn' to escape Earthly gravity you would have to travel at six thousand miles per hour faster.

What if you repeated the phrase again, "one thousand" and you have covered a distance of 186 times the length of Britain? You would then be travelling somewhere in the region of the speed of light. To get to our nearest star neighbour, you would have to maintain the velocity for every second of every one of four and a half years. It is not surprising that our circuits begin to 'smoke' when we try to comprehend distances of billions of light years of space, filled with countless galaxies, some larger and some smaller than the 100,000 light year width of our own, that seem to be all flying out in this massive explosion we are caught up in, and living through. It is not only the galaxies that differ markedly in size but the billions of other suns they are comprised of.

We look up at the Sun and shield our eyes. This huge ball of seething energy. Imagine looking up at one that, instead of being 864,000 miles across, was 400 times larger. Such suns do exist. On a planet there, as near as Earth is to our Sun, the whole sky would be filled with its image. Although our Sun seems to be a single star, many other suns are binaries or triple arrangements. Some astronomers theorise that our Sun may have a dead companion that collapsed violently into a black hole long ago, and orbits some one hundred billion miles beyond Neptune and they theorise that our system was once an active binary arrangement.

Some NASA astronomers also agree with the theory, and searches exist with the quest to find a tenth planet (surmised to exist), indicate with these discoveries, some kind of disturbance to the solar system out towards the 'Oort Cloud' or cometary zone, and that this unseen companion, aptly named 'Nemesis', or Death Star, disturbs the cometary cloud and sends a few of them hurtling in towards us, and has been responsible for Earthly holocaust and the wipe-out of most of the surface life forms in the past, but this may not include the dinosaur extinction as an object with far more mass than a loosely aggregated comet would have been responsible for that E.L.E.

It seems that the threat of the so far uncharted asteroids are not enough to contend with. We have to have a 'Death Star ' as well, but for all that, the dinosaurs did exist for nearly 200 million years, so perhaps we can take heart from that fact. If we exist for that long we should be clever enough to take care of any eventuality, if we are still around, we should be far from our world perhaps despoiling and pillaging the resources of another world in a system far, far away.

To return to the question of fuelling our interstellar spacecraft, the Moon is rich in Helium 3, and this , fused together with Deuterium (heavy water hydrogen isotope), a single tone of it is said to be able to meet the power needs of the entire United States of America for one year. In this regard, it would not be surprising that alien hardware (as has been seriously suggested) is on the Moon doing all that mining out and working the crater rims. It would appear that the sooner we get there and do a bit of mining out ourselves, the better, before it is all gone. I have related to this in my *When the Moon Came* that deals with lunar mysteries and anomalies.

Astronomers, NASA scientists, physicists and astrophysicists including exobiologists, are far from being united in opinion with regard to things constantly being discovered, especially with the enhanced Hubble orbiting telescope. There is no exact science in things celestial our ideas and assumptions are often discarded or modified.

With regard to the still unanswered questions of the Moon rock analysis and the Viking Lander experiments on Mars, it would appear that the sooner we have a base on one of these bodies the better, but then one supposes there will always be conjecture and disbelief rather than agreement in things celestial. Mark Twain's cynical comment comes to mind (perhaps not exactly quoted), "Research has thrown new darkness on the subject and if it continues we shall soon know nothing about it at all". Our probes on their planetary round trips ensure, along with the Hubble telescope, the continual rewriting of the planetary information textbooks. As said, it has been suggested that the Apollo missions were discontinued because an alien presence was discovered on the Moon and that it is the biggest cover-up since the well-known Roswell incident, but even that cannot be said to be proven one way or the other.

If alien beings did make themselves known to us and could solve all these problems and more for us at a stroke, then it would certainly be well worth the 'cultural shock' that their revelations of their presence would cause. It would be a fair exchange if they could do all these things and solve all the problems of starvation in the world and introduce hurricane and general weather control in the volatile regions, but the biggest problem for them would be whether they could trust us.

It is strange today (though not so much as in the fifties), that societies exist that firmly believe, and state, that their members are in contact with extra terrestrial beings that are guiding mankind, etc., then cast doubt on it by saying they come from 'Venus', which is a bit like saying they come from Hell, given the conditions that science, in the form of a softly landed probe, tell us exists there.   The Aetherius Society state, quite clearly, that Jesus was an astronaut and came from the planet Venus. That Jesus was an astronaut has been widely speculated upon, and one cannot blame religious people for being affronted by this and looking upon such statements as sacrilegious. Others, however, with little religious leanings, will say it could not be clearer, (apart from the suggestion regarding Venus).

A 'moving star', an implant by beings in a bright light of a righteous surrogate mother, producing a 'wonder child' who could raise the dead and had wondrous powers, who could not be killed by barbarous crucifixion, was brought back to life, appeared again, warned people not to come near him, rose 'up in the air'. Heaven is, after all, more a state of mind. The only thing 'up there' is space. The Bible tells us, "And when He had spoken these things, while they beheld, He was taken up; and a cloud received Him out of their sight". "And while they looked steadfastly toward 'Heaven' as He went up, behold, two 'men' stood in white apparel, which also said "Ye men of Galilee, why stand ye gazing up to Heaven? This same Jesus which is taken from you into Heaven shall so come in like manner as ye have seen him go". (Acts I, 6-12 and Matthew XXIV, 36). Will the Nostradamus prophecy of 'a great king come from the sky' be perhaps (as foretold), the return of the space entity Jesus? The word 'king' was used in the context of a person of great power.

The above biblical quote is interesting in the sense that it states "This same Jesus that is 'taken'", in other words another group in some craft in the sky was the force that 'took' him. Possibly the same craft that moved and shone down a light for the M.A.G.I to identify Jesus' birthplace.

It is a pity that during Jesus' teachings, he did not enlarge a little more on the statement "My Kingdom is not of this Earth", and of the population of other worlds in general. By now we would be all well-used to the idea of extra terrestrial civilisations and may be spared all that social disorientation and cultural shock, that may ultimately descend upon us, and could in fact be looking forward to meeting our celestial counterparts if Jesus had removed the fear of such an encounter.

We have a stereotyped image of Jesus as a kindly looking being, with long hair, a moustache and beard and long robes, and this is how we would expect him to look if we imagined him coming amongst us today. But, after all, everyone looked like the above in the Middle East in his day, except the Roman soldiers of occupation who, like today, shaved and had their hair short as in the Roman way. So those people in Jesus' day saw him in the same way as we would an evangelist gospeller trying to gather a crown in Hyde Park, wearing a suit, collar and tie (and there are plenty of them). To be sure, the moment one of them starts turning water into wine and working miracles to feed all the world's poor, will be the moment we will sit up and take notice of him, but so far this has not happened. Words are plentiful but insufficient for the human and with all the clever illusionists causing our cynicism, but any entity that could walk on water today, or raise the dead, would most certainly get our attention. If we assume that Jesus was a spaceman, as some written works imply, that means there must be a whole race of them somewhere in space. Who would complain if such a race came among us? who would want to resist? The problems of the world would disappear at a stroke.

If our radio communications have invited such beings into our Earth's space it was the best thing we ever did. On the other hand, if we assume that the same representative of such a race was responsible for the flood, Sodom and Gomorra, the plague of boils and so forth, this is also good enough reason to listen up, as during biblical times humans where treated very much like disposable 'property'

When Jesus said to Pontius Pilate, his initial interrogator, that 'His Kingdom was not of this Earth', it is strange that Pilate did not push this a little further. The statement was quite profound and should have intrigued him greatly, as an educated Roman he must have been aware of all the profound statements of the ancient Greeks, when postulating on other worlds.

We have so much detail on the arrest in the Garden of Gethsemane and the actions and denials of Peter, then the trials and tribulations of Jesus, yet nothing about this other world 'Kingdom'. Educated Romans copied Hellenistic styles in clothing, architecture and manner and must have read the rather advanced writings and theories of the Greeks. It would appear the feeling was not mutual and the Greeks viewed the Romans as somewhat barbaric. Nevertheless, one would have expected Pilate to have led Jesus outside and said to Jesus something to the effect of "Behold, the stars … tell us from whence you came, do other beings really exist in the firmament like yourself? Will we ever meet them?".

With regard to the Aetherius Society believing Jesus was an astronaut and came from the Planet Venus, that may have been all very well when all we knew was the Venus was similar to Earth and had clouds, but they Still maintain it after the scientific discoveries of the conditions existing there, which is strange, but for all we know the conditions we are said to be creating on Earth, ancient humans could have created them on Venus and would have had to depart that world.

If Jesus was not talking about heaven as religion surmises, it is a great pity that he did not enlarge on where his Kingdom or planet was. As said, we would have all become quite used to the idea of extra terrestrial intelligence by now, and his teachings would be no less valid for all that, and it would be a great deal easier for some to believe that, than the concept of a mythical heaven, and the Genesis account of creation which stretches one's imagination to its limits.

When I use the term 'we', I mean those members of the human race that have reported every shape, size and description of alien entities imaginable from Tom Thumb, pixies, Leprechauns and little people, to hairy, long armed, big-eared Yetis, Bit Foots, and 'Giants in the Earth', not to mention the sizes and variety of their craft. It is really not surprising that the debunkers have a field day assigning such reports to the waste paper basket. However, the hard-core of sensible accurate reports tracked on ground radar, air radar installed in fighter aircraft at one and the same time, together with visual sightings by ground and air force personnel simultaneously – they will not go away, and warrant further serious investigation rather than simply being logged filed and locked away.

There must be much footage of pilots' gun camera shots under wraps of weaving UFOs that disappear in front of their eyes in the chase, that could be shown to the public, but probably never will be in spite of any Freedom of Information Acts, but the time has surely come to commence preparation of the public for the acceptance of at least the possibility of an alien presence in our Earth space, and to contemplate and discuss what methods best to adopt other than futile retaliation, better to focus on what we could learn from them.

If we accept all the sightings that come over as unexplained and convincing, and the amount of time they have been seen, it must be clear by now that their intentions, though maybe not strictly honourable, are more benign than hostile, for no other reason that this very lengthy observation of our planet. It is extremely difficult for the human to shed his fear of the unknown and veer away from violence and hostility simply because it is fed to us with a continual diet of such behaviour in films, each more prolific and savage than the last. Children are brought up in such an atmosphere from 'cowboys and Indians', 'cops and robbers' and countless video games containing nothing but violence. How can we shed such conditioning at a stroke or by fine words. Humans have retained such behaviour before and after the sermon on the mount 2,000 years ago and by and large disregarded the ten commandments up to this day.

With regard to the very beginnings of our own human origins, the 'jumps' in the human fossil record do seem to suggest a productive 'experiment' going on. The different types that appeared do not seem to be linked to each other by fossil finds, which makes it difficult to accept a natural primate ancestry but rather a purposeful scientific experiment.

Is the time approaching when it may be seen necessary to do some further genetic 'adjustment' to rid our brains of negative forces and, to stimulate into activity, the rest of our brain cells that may eliminate the negative and destructive tendencies? Is this what the abduction plan is really about? We are moving into their domain now and the alarm bells may be ringing.

When we learn of our past behaviour, such as in pagan Rome enjoying the scenes of Christians being pulled apart by wild beasts, we feel a sense of horror. Similarly, when we see films of the Nazi holocaust, a mere 75 years ago, and the never ending succession of wars somewhere on the globe, so we cannot say we have changed very much over the centuries it would seem that ingrained violent tendencies will always be part of the makeup of some humans on Earth.

We still enjoy it all on the screen and react quite probably to the more gratuitously violent films, just as the audiences in pagan Rome did to live slaughter and, if we could see our faces as we witness it all, the stimulated sadistic gleam would be noticed in the eye of the beholder.

The same applies, of course, to our reactions in boxing and wrestling matches, with people standing and willing one fighter to beat the other senseless.

In the fantasy film, *Westworld*, the lifelike entities provided entertainment for the paying customers of a virtual reality fantasy zone, where one could shoot them in make-believe western or slay them in a Roman arena. If the created entities reacted against humans, they would be removed, their 'brains' opened and tinkered with. Will this happen to us, the human created entities who, with our continuing negative violent acts, malfunctioning and must be seen to be in need of some form of corrective treatment by the Master's representative, i.e. the alleged alien descendants in our midst? Who after all, may have a responsibility similar to the robotic 'creators' and their faulty entities.

We, ourselves would most certainly do this by corrective genetic means if, far in the future we had seen it as our duty to promote intelligence by stimulating some creatures on a far off world who may (for instance) be unwittingly starving to death with foodstuff all around them, or be in danger of becoming extinct, simply due to being feeble-minded. We would carry out the necessary stimulation to their neurones, if we had the ability, then go away feeling quite proud of ourselves about it all. Our hypothetical aliens may have arrived, intending to impart these revelations, long ago. The fact that they have not may be entirely due to the said behaviour patterns, and they may be choosing a much more subtle approach by dealing with the problems by mental insertions in the ongoing abductions.

'They' may come from a society where no one has murdered another being for a thousand years. 'They' may be totally unfamiliar with locks, bolts and safes, and landing on Earth for them, without this necessary observation and analysis, would be like our going for a swim with a school of Piranhas. If they have been in earth space since biblical times, they would know us better than we know ourselves and it would be impossible to assume they are hostile to humans.

In our past, newly discovered races did not originally react with fear to a visiting advanced race, but only with curiosity, even though it was eventually to the detriment of the less advanced race, and it was only later, when the less advanced race saw what the more advanced were capable of that their savagery and hostility came to the fore, particularly if their numbers where being decimated by unfamiliar diseases, and they were being maltreated and their resources looted.

A few examples are the North American Indians and the South American cultures and others still under threat, such as the Brazilian tribes, the Aborigine and Maoris, where originally curious and wonderment ensued until it was too late and their culture began to dissipate and old ways started going for good. It is this type of thinking that prevails in regard to possible alien entities coming amongst us. If they have bridged interstellar gulfs, we could naturally assume that their mental development is equally advanced but it may be unwise to assume anything about an 'unknown quantity'.

Those that seriously report alien encounters that have shaken them, usually finish up with nothing but derision for their trouble, or are seen as slightly odd. One can understand how Hiawatha felt after being ridiculed by his tribe after reporting his 'vision' – "I have seen a big sea water and a great canoe with pinions, bigger than a grove of pine trees and men with white faces and with barking sticks of thunder. Many more came behind them. All our people were scattered all forgetful of our councils. In our forests rang their axes, in our prairies smoked their cities and these men with skin so fair, drove the remnants of our people ever westwards, until wilder and wilder grew the west once was ours". Perhaps such prophetic warnings as these where stated by the 'Mayan' elders but not heeded.

And so, with the knowledge still fresh in our minds of our past acts of subjugation of other lands and cultures and dispersal of other races, together with mass media propaganda in the form of 'B' grade science fiction movies, and bug-eyed monsters and general UFO hysteria with periodic 'flaps', it is ingrained in us to view possible alien visitors as wishing to subdue the Earth, or harm or enslave us. We always return to the inborn fear of the unknown deep in the human psyche, and remove the fear by retaliation and destruction of the threat (if possible).

However, in the absence of any evidence of whether any possible ET entities are hostile or benign, it is perhaps better to opt for the former rather than the latter, which would add up to complete capitulation before the fight had even started.

We can scientifically state that life does permeate space, or at least part of the jigsaw. More and more molecular groups are being detected that make up the matter of living things and it is natural to assume, from this and certain experiments, that life particles are being generated there. Science has artificially reproduced the process going on 'out there'. Chemists have simulated the conditions existing in space by freezing an admixture of methane, ammonia and water to a certain sub-zero temperature and exposing the frozen substance to a proton stream in a cyclotron, and they have obtained acetamide, acetone and urea, the basic building blocks of organic matter. When one considers the possible 95 billion sun-like stars, estimated to exist, some planets around them must have Earth-like conditions, and the masses of protons from those solar winds must be busily producing the same molecular groups in their space of the galaxy and the lightning bolts stoking the seas of those planets industriously producing all the necessary amino acids for life to get started, must surely be going on there, of course it could take many divergent paths, as is evident in the wide variety on Earth, but it may also produce beings with the accoutrements and intelligence for creativity.

Although we ourselves postulate on interstellar flight taking place with humans travelling between the stars, it will be a long time yet. We still have much to do in our own backyard. We are still puzzled by our own Moon. Analysis of the Lunar rocks has only highlighted the mystery, but no doubt, the answers will be found given enough time. We can do little about the inborn nature of the human, to question, explore and expand our knowledge.

We still have two alternative planets in the habitability zone to choose from to create a second or even third home for ourselves if either mineral depletion or population explosions threaten our existence on Earth, but not, of course, a solar threat, although our star seems stable. It has been seriously suggested that our Sun may have already 'switched off' its nuclear furnace based on the absence, or distinct shortage of, certain radiated particles. If it has, it will still be a very, very long time before we would even be aware of it, so perhaps we had better head for Venus if it is going to get cooler, rather like an old gent with no more coal in the cellar, drawing closer to the dying fire.

Stars die, stars are born, and the process goes on. As it stands, there does not seem to be enough mass in the universe to arrest the galaxies' outward flight, and they may go on forever until we can no longer detect them and new life from the huge gas clouds just keeps on repeating itself, over and over again. We can speculate but not prove what the so-called 'dark matter' is all about, but an unknown force is causing the universe to increase its speed of expansion when one would expect it to be slowing down.

As previously stated, when one considers the vast amount of protons in cosmic radiation, it is natural to conclude that basic life particles are being produced in outer space, and they include hydroxyl groups of hydrogen and oxygen and formaldehyde, an organic substance found in the cells of plants. All these substances need is a womb, in the shape of a temperate planetary environment, with ample liquid, and 70% in the case of the Earth with regard to liquid, is ample indeed. Surely, we must conclude that life has arisen on Earth-like worlds and, if there is only one in the exosphere of all the sun-like stars in our galaxy that we calculate to exist, that means 95 billion Earth-like worlds. Even if we halve the figure, it is still a substantial amount. I said in another work that we know for sure that intelligent life exists in space …. ourselves. We are proof of this simple fact. Our Sun would just be another twinkling star to an unearthly observer.

The brain of humans advanced light years ahead of our so-called cousins the chimps, because of enquiring intellect, unlike the apes who stared at something for the brief spell its attention span allowed. We would make the necessary connections after seeing a lightning bolt strike the brown dust exposed from iron ore deposits through erosion. We would pick up the still hot globules of iron and be well on our way to our first smelting facility and weapon manufacture. Whereas the ape may glance at a rainbow for a second or so and lose interest, the human would make the connection with water droplets and rain since they appeared after every shower. After his discovery of fire and what it could produce for him with its heat processes, glass comes, and before long he has a prism and begins to slowly, but surely, produce his lenses and understand the nature of light and takes the first steps towards the manufacture of the radio telescope.

Therefore, in an Earth-like world, where creatures with the necessary intellect exist, it is fairly logical to assume that the advances in science and knowledge we have made simply due to watching, thinking and evaluating, will have also occurred in those worlds. If they occurred millions of years before they arose on our world, then we should expect superior advanced entities to exist that have gone through our process long ago, and all the things we envisage as possible in the future 'they' may have achieved and surpassed in previous years on their world.

As brilliant, an exercise as the Gemini Mariner and Apollo programmes were in getting man to the Moon and returning with (eventually) over 800 lbs of material from another world, they were also primitive, dangerous and wasteful in the extreme. Rockets are inefficient and, with the majority of it falling into the sea, the craft were flimsy and dangerous and a heavy reliance was placed on mathematics and computers to get it right or burn up. The astronauts, fully aware of these factors, went away and they were tremendously courageous. This view would be shared by any observing aliens that may have witnessed what whole programme leading up to the development of the Space Shuttle and, in spite of our more baser instincts, they would have a new respect for human pioneering, endeavour and spirit. Given the extremely lengthy period of time that ariel phenomena and interactions with entities 'not of this earth', that the patriarchs where privileged to, suggest more than a passing interest in humanity, these being that is their predecessors, may even be responsible for our very existence. In which case, they would be delighted with our progress on the positive side but more than a little concerned about our behaviour patterns in the negative mode.

Assuming then, that any day we might see huge interstellar craft move into our Earth space, perhaps waiting in geostationary orbit until the panic subsided a little and respective governments have persuaded people over the radio that 'they' only want to talk and it is safe to return from the hills, how would we ultimately react?

How ironic if 'they' sent down Pioneer 10 spacecraft to us with a thank you note for our able assistance in telling them where we were. It is not only our tell tale radio emissions telling the cosmos where we are. We have sent out the information on gold plated plaques and 'sounds of Earth' and so forth, now heading well out of the solar system. It will be a long, long time, however, before these craft encounter another star, it was stated their primary role was only to examine closely the planets and they where hugely successful in this but they are not heading for any particular star system. We know Pioneer 10 is not even on course for a star system and, on its present course, may not encounter one for 10 billion years and to put such information in it seemed a pointless exercise, but it was brilliant in achieving its operational objective by giving us so much planetary information on Jupiter and the outer planets. Although the Pioneer probe could be picked up by approaching alien craft, they would know all about us through our broadcasts before analysing the information on the gold plated plaques, or listening to the Earth sounds.

However, although (as said) we have 'signalled' a long way; it will be a very long time indeed, before the Pioneer craft get that far. In fact, if we are left alone to develop at the present rate, we will probably think twice about them and as said, go out and retrieve them ourselves, by having the ability of near light speed in order to do so.

If alien entities are already here, their patient and prolonged observation could be taking place from the perfect observation platform of our Moon, and planetary observation craft could be shuttling back and forward from Moon to Earth at frequent intervals, conforming with actual Apollo astronauts' reports on their return trips there. It would be a good explanation for the strange lights and other phenomena observed there and a perfect base for the formulation of their plans for us. They would have noted all our large continental areas of very sparse population if it was 'Lebensraum' they had in mind. With the Moon continually facing Earth, they could be quietly analysing all our defence potential, weaknesses and strengths and, possibly working out their designated landing areas and methods of revelation and approach of the human species, which would be the most difficult decision for them but their presence and observations cannot go on forever.

With their studies of Earth, they would have much to conjecture upon. The very presence of our Moon for instance, they would ascertain by comparisons with Earthly material and the Moon rocks, that it was never part of Earth. Perhaps (unlike ourselves) they may be able to compute the necessary configurations of orbital approach and insertion velocities for such a mass to be 'captured' by Earth, they may have even computed the mystical origins of our Moon. In *When the Moon Came*, I offered a theory that may be found interesting to many regarding its origins.

Many close encounters of the third kind have been seen as though taking Earth samples; they would have means of dating samples and would know that some Moon rocks are older than Earth. Perhaps they could account for the extremely high temperatures required to produce the Maria.

A shared common ancestry with extra terrestrials would most certainly explain their long and (now obvious) presence in Earth space, ever since the 'aliens of Abraham'. Once the massive and traumatic effects of their arrival (depending on the pre-conditioning we may have been subject to) began to subside a little, the most important revelation to humanity would be make it clear to us that humans exist in our present form entirely due to the actions of their forebears, and that they therefore have inherited a direct responsibility for all our negative and undesirable criminal tendencies and proclivity for war and aggression.

Naturally, they would highlight on the positive side, the fact that they had given us the great philosopher's geniuses and great minds that have existed through the centuries. A subtle indoctrination by persuasion would begin they would make it clear to humanity that they have the power to rectify these conditions if they accept the fact that by certain manipulations all human first born could be born without these negative genes in their brains. In addition, they would through convincing persuasion to the judiciary that instead of constant pressure being applied to the politicians to build more prisons, they could if following ET guidance empty them and convert them into affordable housing. The lifers and career criminals would be first subjected to their treatment in exchange for their freedom and enhanced qualities in their mental makeup would increase their chances of gaining suitable employment. If all these things come to pass we could then concentrate our efforts to become celestial travellers ET could not possibly have visited every planet. There is plenty of real estate left for us to explore. With the help of ET, our power sources for our star-ships would become enormously advanced. ET may also save us a lot of trouble by telling that no suitable planets exist around Proxima Centauri, and that it would be far better perhaps to travel to Tau Ceti or Epsilon Eridani, much more 'sun-like' stars of the single variety, where the chances of finding life may be greatly increased. ET may then decide they had fulfilled their Earthly operations and depart to new horizons. All this of course is the solution we may prefer but it remains in the realm of conjecture and we are no less informed what ETs intentions really are or who they really are.

Many of the UFO observations seem to indicate a northerly course and some authors have suggested they may operate from within the Earth, with an entrance in the northern Polar Regions, or simply remote areas. Such aliens would know that, though humans do approach these regions from time to time; they very quickly go home again and do not remain for any other reason than a bit of flag planting. It has even been suggested that the observed phenomena may be us ourselves that have in the remote future, discovered the secrets of time and dimensional travel and have come back in time to verify our history for their memory banks. Well, if that is that case, they do seem to be taking rather a long time about it.

It does seem strange that with all the recorded data going back practically to when writing or recording began, we have (as far as is known) not one single body, craft or part thereof to produce. Of course, if anything did happen to one of their craft, they would not be likely to leave it lying around, but destroy all trace of it. Therefore, we conclude they must be 100% efficient in the operation and reliability of their craft, and show a clear determination to remain unidentified. However, of course when the military show up on frequent occasions with no insignia on their uniforms and recover downed craft and cart them off telling people to forget it, this does not signify highly advanced ET craft but more likely secret US experiments.

It is hard to envisage a life form that is not dependent on water and a planet with 70% water would be attractive indeed. There is a huge diversity of life forms on Earth and every one, as far as we know needs water for life. Some can do without oxygen – in fact, in some cases oxygen is deadly poison to them.

Some life forms, even though not going down to the water hole to drink, so to speak, do obtain their moisture in other ways. We can desiccate bacteria and send it into space but the cells of the bacteria and the fluid therein are protected by the spore wall and can exist for long periods in this condition but will return to normal function once back in a favourable environment.

There has been a theory put forward now for quite a long time by Fred Hoyle and Chandra Wickeramasingh, that such bacteria exist in comet tails and have reached Earth after periodic cometary visits, but how they resist the heat and friction of entry into the Earth's atmosphere when meteorites (mostly) do not, was not explained. However, periodic meteor showers are the remnants of a comet from long in the past that lost its icy material from continuous circling around the Sun and some rocks may reach Earth.

Of course, alien entities even though having a dependency on water, could have any imaginable body form, but the humanoid configuration does seem to be an ideal shape. One can hardly imagine a heaving blob from a 'B' grade science fiction movie, nimbly constructing a spacecraft. Bipedal form, with manual dexterity 'rules'. It is perfectly possible that the makeup of the gases that we are familiar with on Earth, with our 78% nitrogen and the rest 21% oxygen and one per cent other gases, may not be quite the same on some hypothetical alien world that our alleged visitors may hail from, and one rather frightening thought may be that the depletion, that is said to be occurring with our ozone layer, may be a purposeful operation by alien entities to condition our world to suite them. The entire atmosphere is made up of the above percentages of gasses. With all this talk of carbon footprints and additional carbon gas, perhaps it could be explained what percentages now make up the atmosphere.

This hypothesis may be the reason why so many spacemen-like entities in ancient drawings and on cave walls, all seem to have a kind of space helmet or breathing dome on their heads. Is this the origin of the halo seen on 'special' beings in religious drawings, and looking just like an aforementioned breathing dome? They may come from a world with an atmosphere made up differently from ours, and be able to produce hybrid entities very similar to humans to move about on Earth and do their bidding but the actual creators (in modern parlance called 'greys') would have to remain in the confines of their craft for most of the time but they may be able to tolerate our atmosphere in small doses.

During this lengthy observation of our more base activities from pre-Roman times to the present day, those observations would have included the most monstrous tyranny ever to have occurred on Earth that of the Nazi regime.

The colossal crimes against humanity perpetrated over six years, with millions of people purposely exterminated simply for ideology or beliefs, was quite mind boggling and how so many of them went so calmly to their deaths without reacting, rebelling, panicking or running riot, remains something of a mystery. Today however, it seems that nobody can push them around having settled in their 'promised land', but things were different under Hitler, during the Nazi regime. It was almost as though the victims accepted such treatment as their destiny. It is not surprising that many viewed Hitler as the Anti-Christ on Earth, and this factor makes it even more strange that the Vatican stayed so deafeningly silent during the Nazi regime and its evil actions. In fact, the recognition by the Vatican of them coming to power in 1933 gave Hitler his first greatest prestige.

In fairness, it must be said that many people of the world, apart from hearing certain rumours, were not fully aware of the enormity of the Nazi crimes until after the invasion and the subsequent Trials of Nuremberg.

In fact the US Government assumed that the alleged crimes against the Jews where either Russian propaganda to hasten a second front or compiled by the Jews themselves in order to gain easier access to the US and the West. Although it does not say a lot for humanity in general, the fact is, that the US and Great Britain did not wish to import large numbers of Jews. At the time they where the butt of jokes made in poor taste. Many people seem to envy or dislike their proclivity towards making money but mostly because many people had been instilled since their school days that Jews were responsible for the death of their 'Messiah' Jesus. It was a long time before the truth regarding the Holocaust emerged. But as the allies and the Russians advanced into Germany and discovered the extermination centres and news began to trickle back to the US and Britain, we could say that even then, many denials and accusations of propaganda were made, which is why the Allies who of course included Russia, were so careful to film and document it all and even make the local civilian population witness it all, and not one of them admitted any knowledge of what was going on. In all fairness, due to the secrecy of the operations and also reflected by the shock and horror on the faces of the villagers this may be true. Some females fainted and it was said that a local Mayor committed suicide.

When the Russians liberated the camps in Poland on their advance toward Germany, they invited members of the international press to come and see for themselves, and even then, some would still not believe it. If we ourselves find it hard to accept that human beings could inflict such things on other humans, imagine the horror possible advanced alien entities would have in witnessing such things occurring on Earth. Of course, the horrors did not stop with the Nazis. The whole Pacific scenario would have been witnessed with the Japanese war crimes and the actual use of nuclear weapons in anger, and their horrifying aftermath, that seemed the only way to stop the fanatical 'death wish' to fight the last man condoned by the emperor in his lofty abode and escaped any retribution for war crimes later.

With regard to the apparent silence of the Vatican to speak out or oppose fascist regimes in Europe, the interpretation of the word fascist in the dictionary is simply anti-Communist, and it is clear that the Catholic faith and communism were completely at opposite ends of the spectrum and, furthermore it is quite well-known that there is little love lost between the doctrinal views of the Jewish and Catholic faith. If the Vatican, perhaps in an enclosed, naïve and ecclesiastical way, condoned Hitler's opposition to the godless doctrines of the east and to the Jews, there would be little temptation to verbally oppose them. However, in their, what could be called 'ivory tower' existence, one would not expect them to be aware of the appalling atrocities being committed, so it ….. may be unfair to suggest that the Vatican or anyone else, for that matter, could have been aware of them occurring and to come in man's inhumanity to man in German occupied Europe. One might ask who are we to criticise with certain members of our own royal family appearing to sympathise with the Nazi regime for reasons best known to themselves. Edward the VIII for example who seemed to admire Hitler and was seen 'glad in handing' him the newsreels. Certain members of the aristocracy also admired them. We may recall Oswald Mosley and his black shirts also there was a large Nazi movement in the US. Also when viewing the Vatican as having kept silent Jesus himself was careful to keep outside the political arena during his travels and preaching, and is not recorded in Biblical texts, heavily criticising Roman atrocities.

However, with the benefit of hindsight it is easy now to imagine Hitler as the 'Anti Christ' by establishing his counterfeit kingdom on Earth by pulling Germany out of chaos into order, albeit in an autocratic and totalitarian way with all those fine buildings, employment for all, autobahns and national fervour, but if Hitler was acting in accordance with biblical revelations and prophecies, where is the new and peaceful order said to follow his defeat or Armageddon, but in certain respects he was the great deceiver and most certainly a false prophet.

The lion is supposed to lie down with the lamb and we are supposed to be busily beating all those spears into ploughshares and knowing war no more, yet in a recent remembrance service had a news commentator saying "And many ex-servicemen came to pay their respects to the dead of two world wars and the many conflicts since then". The First World War was said to be the war to end all wars yet we learned nothing from it as fifteen years later the Anti-Christ came to power and ensured that there would be another world conflict. In all fairness, some nations who yearned for peace and stability where 'dragged' into it, yet striving to remain detached from it all. In 1939, Neville Chamberlain did his utmost to keep Britain out but is seen now as having been weak and appeasing with regard to his efforts and history remembers him as such, when in fact he should have received the Nobel Peace Prize for his efforts to maintain peace.

However, for all that, in the long run governments can only really be sustained by the will of the people, and to be sure any country trying hard to maintain such regimes, will one day fail and the only true form of government, that of democracy, will prevail. If there is any better system for the people, we have yet to hear of it. Hong Kong was rightfully returned to China after its 99 year lease but after all that democracy, to find themselves under more a restrictive regime was something of a shock to them and difficult to adjust to.

The Eastern masses seem to be just as easily led as Europeans were in 1933, by a leader possessing the rhetoric and charisma such as Mao Tse Tung, but even if he eventually fell out of favour. Originally he could do no wrong and the next he is unceremoniously disregarded. How are the mighty fallen? Looking at the few, if any, civil rights existing there it would take courage indeed for such a leader to emerge in order to speak out against it all, but perhaps if the prophecy is borne out, like it is said, some of the other Nostradamus quatrains have done, then would this 'leader to be' emerge from the ranks of the exiles? , working at the moment, rather like Charles de Gaulle from Britain during the days of occupied France, perhaps ET has someone in mind, which will be just one of the aims of the alleged alien observers when they finally decide to appear to us, but they certainly would not win the hearts and minds of the people by choosing and inserting our political leaders for us. He or she would have to appear as worthy of the position.

If the biblical characters were observing aliens, 'they' most certainly have had lots of time to formulate a pretty good plan. They have assessed us well and would know Earth's history backwards and forwards. They would know all the world's leaders existing (and possibly to be), their character profiles and political ambitions, and would certainly know who would be 'the best man for the job' so to speak. They would certainly know who to install, where and when to best fit in with their ultimate aims for us. these beings may even be some form of hybrid entity, part extra-terrestrial and part earthling in makeup, perhaps already in existence like 'sleeping agents' awaiting the call.

Although it is firmly entrenched in the human psyche to resist and fight against anyone trying to force their will on us, it is perfectly possible that 'they' (being our alleged observers) may be able to make us an offer we cannot refuse, in that the benefits and advantages to humankind would be enormous, and in this regard we may accept 'them' in the same way as we would a convincing leader or politician offering all the right policies with persuasive rhetoric, but they would have to be very convincing.

Another quite human trait is that we also seem to crave good leadership and someone to 'look up to' in almost hero worship fashion. This is notable in the American psyche ever since the first comic hero Superman and if no one is to hand, we await a mythical 'messiah' to eventually come and deliver us. we seem to require a role model to relate to and to guide us, not only Superman who first appeared in the thirties but a constant stream of the, Batman, Captain Marvel, Captain America and so forth. Will this era (the current millennia) be the dramatic time foreseen by the seers, when our hypothetical aliens will appear? With all the cases on record of alleged abductions perhaps their hybrid entities are all installed in many important posts awaiting the call, like (as said) sleeper agents, for the completion of the final part of their plans for the 'human question'. The more such hybrids that could be inserted into, or achieve, positions of power and influence the easier their task of appearance and control would be.

As said, the well publicised case of the supposed abduction of Betty and Barney Hill was extremely convincing and, if we cannot believe that one, it is hard to believe any of them, but the point is, it occurred sixty years ago and no doubt there were many others that were not 'hypnotically regressed' and considered that their experiences were either dreams or experiences that may also have occurred. The primary task of the hybrids would be to encourage the notion that acceptance rather than resistance should occur and that the advantages in medicine, genetics, science and space technology would be enormous, rather than futile opposition where nothing but destructions and wasteful loss of life would occur. However, for all that, many people would realise, in particular psychiatrists that such acquiescence is totally contrary to the human psyche and not easily bred out.

Even the doctor who carried out the retrogression on the Hills could only decide in the end that the couple had simply shared a dream, in spite of the fact that during the retrogression sittings Barney Hill became so stressed that he had to be quickly brought out of his trance state, possibly for this own safety, but more worrying, other abduction victims have been informed telepathically that their operations are drawing to a close.

There is another very convincing abduction case on record that was alleged to have occurred in Pascagoula, Mississippi. In that particular case, the victims were a Charles Hickson and a Calvin Parker, where the abductors appeared to be more like some form of automaton or robot entity than the usual form of alien reported. This departs considerably from the usual experience that the victims of abduction have experienced, where human looking beings and the smaller so-called 'greys' are usually featured.

Again hypnotic regression techniques were employed and Charles Hickson appeared to relive a very vivid experience of human abduction, with some form of medical examination having taken place in it. Both the men were clearly quite shaken by their experience and when left alone with planted sound recording equipment in the room were heard to pray, and it would appear that their main concern, quite naturally, was whether the alien entities, or their mechanical agents would come back for them or try to find them again. Even if they had suspected that recording equipment was installed, the dialogue still came over as sincere and convincing and the somewhat naïve down to earth conversation was quite in keeping with the two unsophisticated normal people that they were. Clearly if they had invented the whole affair, they would be worried about getting away with it and perhaps joking and expressing how gullible their interrogations were to be believe them.

If such a thing happened, it would be essential to the victims to tell somebody. It would not be an experience one could just go home and forget, but it was clear that in this case, as in the Hills', the victims were not seeking any form of notoriety and seemed, in fact, quite embarrassed by it all. It was the physiological stress of keeping it to themselves that was said to have caused Barney Hills ulcers to flare up, where before, he had seemed to be improving.

Perhaps significantly, there did not seem in the Pascagoula case, to be any attempts made by the perpetrators to make them forget it immediately afterwards, and the significance could just possibly be that the aliens now no longer care whether we know or not about their activities as it is so close, possibly, to their final appearance, that some form of prior pre-conditioning to their presence may actually now seem desirable. Clearly if all the most recent claims of abduction where analysed with regard to the answers some victims get such as "Our operations are drawing to a close", it may indicate a coming traumatic event.

If the abduction cases were confined to classes of people who were known schizophrenics, or from show business outlets, or from groups known to be fond of seeing themselves 'in the limelight', so to speak, there would be grounds for assuming overactive imaginations and publicity were the cause and not alleged aliens at all, but this does not seem to be the case and, if anything, the opposite seems to be true with regard to the character profiles of the 'abductees'. In the Hills' case, perhaps it would have been even more interesting if the hypnotic regression sessions could have been carried out nearer to the actual event, but many victims often have to be coaxed and encouraged to report their experience and any analysis takes place much later, but a well trained hypnotist can draw them out as though a recent event had occurred.

A feature of many of the cases regarding alleged abduction, seems to be the suspicion that something has been implanted in the body of the victims, enabling the abductors to be capable of quite easily finding their victims at some later date. Perhaps a thorough X-ray check would be in order on any future convincing claims of abduction for signs of any small 'implants', as implants have definitely been found in and removed from victims, they appear to have been treated in a manner that would prevent the human body from rejecting them.

Although it appears generally that such things, as abductions are not taken too seriously except for a few highly publicised cases prompting regression, the average man in the street has no real idea just how serious the authorities may take them. It would not be generally known, whether, after all the news media and public interest has died down a little, the 'victims' may have a visit from certain people to quietly gather the facts and swear the 'abductees' to silence with some kind of veiled threat, enter the M.I.Bs or 'Men in Black'.

This kind of activity is usually assigned to the so-called MIBs or Men in Black. The interviews could claim to be government agents with impressive credentials, which may be verifiable at the time, but later on, it may turn out that no one has ever heard of them. The original MIBs first appeared in the 1950s. They differed greatly from their Hollywood image; they had pale skin very red lips and always arrived in black limousines far from new but always appeared shiny and new. They appeared almost robotic, and visited those who had reported UFOs.

If the UFO 'victim' does co-operate then later admits a visitation of government agents, it would all be denied and they would be untraceable and, in the end the 'victims' would be accused simply of trying to arouse further publicity of their case for financial reasons, and would end up receiving more ridicule that anything else for their efforts. It would seem that modern reports of men in black have been improved with regard to their image but one noticeable thing reported is that they do not blink and do not seem to have eyebrows.

The well-documented Travis Walton abduction case that was later made into an interesting film had all the classic ingredients of derision and disbelief, accusations, incredulity, then a division between believers and sceptics. If the events in the film were accurate, it seems a group of forestry workers returning home were alarmed by a bright red light and something hovering in the woods. Mostly, they were quite alarmed and beginning to panic, but Travis Walton was seemingly more curious than frightened, and walked toward the light and when he was directly below it, a bright column of white light seemed to encase him, lift him off his feet and throw him on his back. At this point, the rest of the group became somewhat hysterical and panicked, causing the driver of the pickup to accelerate away from the scene. Later the driver shed his passengers and went back for his friend who, by this time had disappeared. One could not help being sympathetic to them and their problems in trying to convince others of the event and convince them of their innocence.

At first, there was the usual derision and disbelief, even veiled accusations that they had killed their workmate as there had been aggression among them. Five days later the victim turns up shocked, cold and incoherent at a gas station.

After eventually submitting to a lie detector test, which proved inconclusive, there was still disbelief and many assumed it was an attempt at financial gain, as the film indicated a few of them were struggling a little financially.

Although it was not made clear in the film, the events depicted during Mr Walton's five-day absence, gave the impression they had been drawn out by hypnotic regression, and were quite profound. Although 'dream like' the scenes where he was floating and throwing himself about if he made sudden movements, were all-conductive as to how things would have been if he was weightless. In other words in space, or at least in orbit on board some huge craft.

A good description of the alien entities was obviously given unless they were from the imagination of the film makers. It seems Mr Walton became rather enquiring on board the craft and broke out of his flimsy skin type cocoon and began floating about the craft. Then, it appears, the aliens lost a little patience with him and dragged him off by the legs to an 'operating' table and gave him a thorough once-over. All subsequent lie detector tests proved positive in favour of the group's story. However, there is a contradiction here in that where Travis Walton seemed to be weightless the ETs did not.

If it was a hoax, Mr Walton most probably would have simply walked out of the woods after the five days having been secretly fed and looked after by his co-perpetrators during his 'absence' and would simply claim he had been returned. It is hardly likely he would have subjected himself to hypothermia and possible pneumonia, or the cuts and bruises on his person that were apparent from his medical 'examination'. He was also unshaven and appeared as though they had been away from all the comforts of home.

As previously suggested, Mr Walton almost certainly had a few MIBs calling on him when everyone else had got through with him. It is more like human nature to be suspicious and wary than to be gullible and believing everything that was related to them. Otherwise, detectives, for example, would never solve any crimes, they would simply believe the accused and their story.

Clearly, to be in the position of an abductee is not one to be envied, even if they did make a pot of cash from films and books about the event. They would still want to be believed and would probably deserve the cash. If it did take hypnotic regression to uncover the events that occurred in Mr Walton's absence, clearly the alleged alien entities did not wish their activities to become known. However, does this only apply to one particular group? Other abduction reports seem to try to ensure their victims that the ET operations are drawing to a close and may wish it to be made known to the authorities.

Some reported UFO activity seems to suggest others were unidentified purposely tantalising the obvious limitations of Earthly craft when in pursuit of 'blips' or 'bogies' they had been directed toward by ground radar, so obviously out-manoeuvring them, and even winding up on their tail. This happened to an RAF jet in the fifties and the pilot gave up, partly in exasperation and partly being low on fuel, they would most certainly report and believe that some other world technology was responsible for their experience the victim would clearly be very shaken.

A lot of older reports of unidentified craft, which is from the last century, seem to indicate a similarity to the technical state of advancement of their day, i.e. chugging dirigibles with strange people on board. These reports are almost certainly emanations from over-ripe imaginations, but those that are reported in those times speaking of discs and orbs of light travelling at great speed, deserve a little more attention, and most certainly, the 'flying shields' of biblical times, as many shields were of course around.

Of course, the observers struggling to describe what they had seen in even older times, would have no level of aerial technology to draw on and could only look around for something similar to describe the craft, such as flying carpets, but a 'flying shield' does not seem to be anything else than a flying disc, as per those reported today. Ball lightning is nothing new and has been the subject of ancient reports from very perplexed people and never seemed to be confused with flying 'shields'.

It was usually reported as slow moving with streamers coming off it and floating along the ground, eventually exploding and giving off a foul odour, (which it would appear, is how it does behave) but ball lightning is distinctive in its own right and cannot be used to classify all sightings, neither can swamp gas or weather balloons.

Kenneth Arnold, no doubt found his observations in 1947 that coined the 'saucer' description, quite financially rewarding, and who would blame him for not wanting to kill the goose laying all those golden eggs. However, in fairness to him, he did say that he never actually said that they were 'saucer' shaped, only that they skimmed along like a saucer would if thrown across a pond. His actual description was crescent shaped and this, together with the undulating motion, seems almost certainly to suggest a flock of high flying geese, but that would not explain their excessive speed calculated by Arnold as they flew from peak to peak in the mountain range, if they calculated it correctly.

However, there were a lot of reports already in before Mr Arnold's famous sighting. An observation by commander of the USS Supply, Lt. Frank Schofield, together with his crew, saw three huge luminous objects flying about in the South Atlantic and his report came in the year 1904. The longevity of the phenomena is always made clear by this wide range of dates in all the centuries back to biblical times.

In the journal of The Royal Astronomical Society of Canada, a certain Professor Chant of Toronto, reported groups of luminous UFOs flying in groups of two, three and four in perfect line abreast formation along the borders between Canada and the USA. This sighting was dated as 1913. Serving pilots in the RAF and other services of other countries, have an official reporting procedure, which does not come to the ears of the general public, and civil airline pilots are very reticent in reporting such things unless they are so obvious that the entire passenger group also witnessed them, thereby eliminating any suggestion that the pilot was somehow unreliable.

Civilian pilots are only too aware of what happened to some of their counterparts in the past after reporting such things, and then when the time came for the periodic medical check-up found themselves out of a job. It was not viewed as good customer relations for an airline to have its pilots 'seeing things'.

If there are actual alien entities manifesting themselves 'out there', then every human reaction seems to play into their hands. It would suit them that most of us are sceptical. It would suit them that many see and do not report. It would suit them that a certain amount of ridicule and suspicion is heaped on those that do report. There really does not seem any point in hypnotically regressing people in order to get at the real facts if those people faithfully report in graphic detail, all the various aspects of the abduction or encounter and then, as in the case of the Hills, the investigator says they simply dreamed it all. He might just as well have said that in the first place and saved his effort and the wear and tear on his equipment running the tests, but of course, there was always a fee to be paid, and his task was simply to relieve the victims of their trauma and anxiety.

Clearly, in those particular cases, it all seems too incredible to believe, which again falls neatly into the alleged alien observers' favour, and the operators of hypnotic regression and lie detection techniques are well aware, before they start, that the methods only reveal what the 'victims' believe to be true.

One unfortunate pilot was never given the chance to speak or be tested in any way as he totally disappeared during his encounter in 1978 in a light aircraft while flying to King Island from Melbourne in Australia.

He and his craft totally disappeared after a huge UFO hovered over him and his Cessna 182 over the Bass Straight, and there was a conversation between the Melbourne Flight Service and the pilot on the evening of October 21st 1978 at some time after 7 O'clock. He reports a dark circular shape descending over him. His last words were …….. "That strange aircraft is hovering on top of me again …. And it's not an aircraft". Apparently, there was a strange metallic sound and the conversation was terminated and after that last message pilot and aircraft totally disappeared, this seems to strongly indicate that the UFO swallowed up the aircraft and it is now on display in some ET museum.

At the time, a UFO was reported by various sources and was described in one version as having lights at the extremities of a 'starfish' configuration. The film *Close Encounters* was said to have drawn heavily on actual reports and happenings and the description of the craft in the above case brings that type of craft to mind. When we mentioned the above case quite a number of ships and aircraft have disappeared, are they now items on display in some ET museum on a planet far, far away? Five Grumman avengers from the missing flight 19 and other aircraft for example have disappeared in the Bermuda Triangle area, also any number of boats from the same zone, but we are brought down to earth again with the laconic remark by the late Professor Carl Sagan when he asked "Why don't trains disappear?. Nevertheless, we must ask is the reason the occupants are never returned only because ET does not wish any aspects of their methods and technology to be relayed to earthly science? Or is there a completely different reason. There is a strange analogy in the cases of N.D.E or near death experiences, that are more numerous than people think. During operations, the patients under surgery sometimes appear to pass away and a desperate fight takes place to resuscitate them ensues. During this time the patient seems to pass through a tunnel and emerges into a warm loving light being welcomed by lost loved ones and does not wish to return until some 'voice' says "You must", the patient then quickly recovers his spirit back in his body.

Chapter 4

DO WE HAVE COSMIC ANCESTRY?

It has been suggested from various sources that human kind are 'children of the stars'. Well, in a sense this is true. Almost everyone knows that all things on Earth, indeed in the entire universe, are made of atoms. Every atom in our bodies and in the bodies of all living things, and in everything we see around us, was once produced inside a star and was flung out into the universe, enabling other stars and planets (that we know exist around other suns), and the lifeforms on them to evolve. The dust of the Earth, indeed all its planetary material is 'star stuff', so the funeral expression 'dust to dust' makes sense.

Any alien entities that may be discovered by our future space adventures or appear at some time in our skies, would also be made of this same 'star stuff', although not necessarily in the same arrangement as ourselves. Nevertheless, made up from the same list of elements that we hope we have totally discovered. Although the demise of the dinosaurs has a cosmic signature who is to say that it was not intelligently instigated and later any remnants hunted to extinction, thereby explaining the ancient boot prints found in stone.

Since the whole Darwinian concept of human evolution in the miraculously short time of modern human existence is questionable and certainly did not involve millions of years ago. For all the fossil hunting, it remains a theory and riddled with conjecture, it is perfectly possible to consider that humankind may have a 'cosmic' ancestry, in that we may have arrived on Earth as cosmic 'refugees' long ago and possibly degenerated and then after sinking into a semi-savage state, due to many hostile planetary conditions before gradually recovering. However, in that case human bones would be found from far in the past but remain the same and would show no signs of gradual development .

Human bones have been found along with hominid fossils, such as 'Java man', proving we co-existed alongside them in the past.

When we contemplate the possibility of human entities as being 'produced', so to speak, by hypothesised alien beings carrying out some scientific creation process, it is largely encouraged by the circumstantial evidence of the 'jerky' procession of 'homo' types that 'suddenly' appear in the fossil records after the hominids with no links between types. This is the problem that so confounds the anthropologists. Perhaps their hypothetical entity, the 'pro-consul' did exist in spite of there being no fossils to verify him and he lived alongside the early hominids, and his early work went back as far as the true Pongid apes, but to suddenly breakaway from them to be a distant human ancestor would have been a monumental event that defies all known anthropological laws.

Some anthropologists have suggested that the 'branching off', or mutation, could go back this far. Blaming simian types for the negative forces that retard our advancement and produce the savage manifestations in humans, such as our proclivity for war and our worldwide crime rates, may be a little unfair to them, and could well be considered as the result of a not quite fully successful genetic experiment to enhance intelligence, with the different chemistries involved possibly reacting with each other.

The period of fifty million years ago, in being so close to the termination of the dinosaur rule, could have perhaps seen the evolution of a gradually large primate from the smaller varieties that obviously survived the extinction event, particularly as this tendency for largesse existed in their genetic makeup, as indeed did bipedalism, but even though Tyrannosaurus Rex existed for millions of years, bipedalism counted for nothing in terms of his development.

This creature would seem to be a much more fitting candidate to blame for our still retained and well manifested behaviour patterns after all, even though apes and gorillas can be aggressive, they are not inordinately so, and gorillas for the most part only appear savage in their body form and resist going to war with each other in droves (that we call armies) and are fairly peaceful creatures, but referring back to the Dinosaur Era, most species were plant eaters, but not T Rex, he was a savage bipedal carnivore as indeed (one might cynically remark) are many humans.

Certain anomalies exist that science has quietly ignored, but which will not go away, a few examples of which are, the fossilised human brain found in Russia, dating from the Carboniferous period, the apparently 'machined' cube of polished steel measuring 70 x 70 x 50 mm with a groove cut around the centre, which was found in a coal stratum in Austria. A full report of this find is contained in "Nature", 11th November 1886, page 36. Clearly, it was 'machine' produced at a time when we had no machines.

There is the unexplained shoe, or boot print, found in ancient sandstone from millions of years ago, found in the Gobi Desert in 1959 by a joint Soviet/Chinese Paleontological expedition there. It has rows of indentations across it, about seven to each row, and ridges running the full length of the print. It reminded the discoverers of a climbing boot print.

Many legends of almost all of Earth's peoples speak of 'sky gods' and 'creators'. Some archaic and ancient Indian texts have many references to such beings. The Hindus referred to the ancient beings as 'Siddhas' or 'Possessors' and 'Masters of High Science'. The Hrussos or Akas of India's North East

Frontier say that at one period in prehistory, there were no men on the Earth and that we are 'all descended from space beings that 'migrated' here in antiquity'. What can we make of ancient texts speaking of 'Vimanas' or flying machines and descriptions of atomic explosions and missiles? Also, tales of the birds turning white, human hair falling out and elephants blown away like leaves.

Eastern Asian legends related that Japanese and Chinese people's were settled and civilised more than 1,000,000 years ago. It would appear that a form of civilisation existed much further back in time indicated by an apparently 'tiled' pavement discovered in Plateau Valley, Colorado and dated at 12-26 million years old. How difficult is it to imagine ancient workers laying tiles, so long ago we naturally suspect our dating process and their accuracy, because such 'ancientness' with regard to humans is so hard to accept with no ancient bones matching such an age. The age factors should not be important since we can find 65 million year dinosaur bones.

Then we have evolutionary theories with scant fossil evidence and that totally ignore the unexplained enigma of the human brain and how it could possibly have achieved its enormous intellectual capacity in such an extremely short time period, when plants, creatures and insects show such genetic stability for a massive amount of development time. It can be seen, therefore, that the 'extra-terrestrial hypothesis' is not so fantastic after all. Everything appears to suggest some form of interference or manipulation during the evolutionary process.

Nevertheless, certain people will reject it out of hand, but will accept the fact of science telling us that in the future we will achieve fantastic things, from molecular and atomic 'restructuring' to amazing genetic manipulation and experimentation, by 'creating' creatures to order and eliminating negative and destructive neurones and enhancing intellect and suchlike, will be commonplace by humans on humans, they will accept 'wormholes' in space, anti-universe, tachyons time warps and way out scientific postulations as acceptable simply because they have been proposed by science, and are said to be mathematically feasible. Many would view it has human arrogance to assume we are the only beings who will be capable of such things.

We are today, looking hard for reasons to explain insanity and mental disorders. We are not looking just to satisfy curiosity, we are looking to find cures and those cures may ultimately be neurosurgery and genetic 'adjustment'. If we consider alien entities with some half a million years of advancement on us viewing human behaviour patterns, they would be well aware of our shortcomings, yet *they would know we had plans for cosmic travel.* Left to our own resources, it would appear that humanity would have war upon war until the end of time. They may decide that as the future King Edward VIII said when visiting poverty-stricken areas also ravaged with disease "something will have to be done".

They may see contact with us as totally out of the question and be already carrying out certain experiments with all the convincing abduction cases we hear about in this regard, for a master plan to solve 'the human question'. Could it possibly be that it was alien entities that actually enhanced the intellect in us in the first place? In addition, the alleged aliens are here to carry out the final adjustments or 'fine tuning' before seeing us as being ready to take up membership in the galactic organisation of advanced life forms, but firstly to enlighten us with their 'revelations?'

It seems inevitable that we will eventually develop a suitable power source for interstellar travel and we will set off for the stars. Who knows what wonders the occupants of such craft will behold. Future astronauts will probably not be majors, colonels or commanders. It largely happened that way in the commencement of our space activities due to the natural flow of such people from the armed services to becoming test pilots, and eventually astronauts in going on to greater challenges.

The astronauts of the future (and the process has already started with the Space Shuttle) will be doctors, professors and various highly qualified people, male and female, in engineering, electronics, robotics, geology, exobiology and genetics and all the branches of technology considered necessary to make the voyages, where discovery of extra terrestrial life forms may be encountered, and any visiting aliens we may have in our Earth space would also be quite proficient in these fields.

If in some future time, astronauts from Earth encountered a species, in some far off system, becoming extinct through its own underdeveloped senses and lacking the intellect to save themselves, by using that which may exist all around them, in perhaps, a food source they never imagined. The astronauts may be sufficiently clever to probe into their brains, and enhance the intellect of such creatures by stimulating the necessary neurones to allow them to save themselves.

If our imaginary aliens were masters of levitation, ESP, mind control and telepathic powers and great mental ability, and we have inherited some of this genetic material in our brains, perhaps this is the reason that many seem to display flashes of or can actually manifest these powers, though what some people can be observed to do under controlled conditions can be quite successfully imitated by competent magicians, but there are those that claim their power are genuine, Uri Geller for example. He chooses to prefer an almighty God as being responsible for his powers, but what process would exist even for an almighty God, in his grand exalted level, to choose people like Uri

Geller for his endowments and others to be imbeciles or grossly deformed? What deciding factors would influence His choice, or would He just adopt a 'You, you and you' policy, like a parade ground drill sergeant picking out his 'volunteers'. One could not 'aspire' to be chosen for greatness by one's deeds as humans are already born with their gifts or afflictions as the case may be.

Another example of special power is clairvoyance, and also there is the ability of certain people to handle mentally, and with great speed, enough to beat computers many lengthy calculations and computations in their head, and, with the use of the simple abacus or bead system. Other strange cases see certain humans appearing half mentally retarded yet possessing great ability in one specific thing, such as the aforesaid mental arithmetic ability. War and conflagration have been responsible for the retarding effects on the advancement of humankind in an unbroken line from the Cro-Magnon intelligence of many thousands of years ago and, indeed, many of the achievements of that entity, the direct ancestor of modern man, puzzle us even today. As well as their advanced and well known cave art, they were proficient at clothing manufacture and precisely manufactured artefacts, and they even had a mysterious process of being able to straighten Mammoth tusks. Spears have been found to be manufactured from them.

When contemplating on some of the puzzling artefacts of the Cro-Magnon, there is an artefact called the 'Baton of Montgaudier', a 370 mm long piece of reindeer antler that could only have had its fine miniature engravings carried out by the use of a jeweller's magnifying glass. If they had magnifying glasses, one could assume they had telescopes. Alexander Marshack, one time Research Associate of Harvard University's Peabody Museum, established that, 34,000 years ago, Cro-Magnon was keeping track of Lunar cycles by the use of abstract symbols. Everything seems to point to the 'giant leap' of advancement and intellect in the human entity as being somewhere between Neanderthalensis, Sapiens and Cro-Magnon, i.e. some 50,000 years ago. Are we today just relearning things we knew and took for granted in the past? Just as refugees from an advanced world would degenerate with the sudden loss of their technology, so would a race subject to a geological catastrophe where a few survivors prevailed and moved on to more arid or hostile regions, making survival that much harder and a return to basics once more, then having to relearn everything again as children, just as was said to Solon by an ancient Egyptian scribe when stating previous Earth catastrophes had happened in the past.

If we had been the victims of the series of cometary fragments that battered Jupiter, all our civilisations and most of our populations would have been obliterated, BUT, pockets of humanity would survive somewhere and, to be sure, those of humanity we now look upon as primitives, such as the Amazonian tribesmen, natives of New Guineas, Aborigines, Maoris and Eskimos, would have their share of survivors and these beings would be the immediate new masters of the world, being instinctively better equipped to survive and any knowledge, except for legends of previous glory, would all disappear with the rubble of the wreckage of civilisation everywhere and the slow process of the human entity, equipped only with his intellect and racial memories, would begin the slow march once more toward progress.

Chapter 5

THE ULTIMATE PURPOSE?

O ur thoughts must go out to any person who has been the victim of an abduction, whether it be imaginary to an actual occurrence. If it was an actual occurrence, the implications are quite fantastic and not many people would wish to dwell on them for too long, but we should be prepared to face the possibility that they have a profound and ultimate purpose. The trauma to the victim would be no less severe if they firmly believed it had happened, but to ease their mind submitted to hypnotic regression sessions, for the simple reason that if they did firmly believe it, then this belief would manifest itself during the therapy and be horrifyingly confirmed to them, but at least they would know that they were not losing their sanity, but it would not ease their trauma.

Dreams can sometimes be confused with reality. If one began to suffer flashbacks and repetitive dreams and this was accompanied by lost time that could not be accounted for, such as in the Betty and Barney Hill case, this would most certainly reinforce the belief in the alleged victims that it had happened just as they believed. If they had fallen asleep in their car, they would have been well aware of that. They would remember waking and feeling chilly and waiting until the head cleared before driving on again. If the Hills had fallen asleep then it would have been easier to accept it all as a dream, but they were adamant they had not been asleep any time during their journey, and the possibility that they may have dreamt it all was simply due to the 'lost' period of time, and the hypnotic regression session clearly indicated that the lost time occurred during an abduction.

The cases where one believes that their abduction took place while they were asleep are the difficult ones. It would indeed be an unenviable position to be in, for their trauma would be more severe, as it would be extremely unlikely that anyone would believe them, including the therapist conducting any hypnotic regression sessions, it would always be assumed the victim had been dreaming.

As for the therapists, they would make the assumption that what they were doing was in effect, simply bringing up from the depths of the patient's mind, what he or she 'believed' to be true. It is a feature of the human brain to repress anything too unpleasant that one would not wish to dwell on. This is just the mind's attempt to protect itself from stressful thoughts, but to be sure, an alleged 'victim', finally driven to the point of, probably expensive, hypnotic regressions to gain some peace of mind, would surely not get that peace of mind if the results of the tests just seemed to confirm their fears that it did occur as they imagined it, they would find it very difficult just to simply forget it and get on with their lives.

The therapist would, as said, assume their patient had dreamt it all, but would the patient be really convinced? Strong hypnotic sessions to assure them had been 'all in the mind' would be necessary. In any case, the alleged 'victim' would never be able to look up at the night sky again with the same feelings. They would always remember it and somewhere in the back of the mind would be the nagging doubt that 'aliens' would possibly revisit them. On top of this would be the unsought publicity, finger pointing and the general feeling that there must be something odd about the victim to have stated such things. Betty Hill communicated with the aliens verbally and they stated (not verbally) that they could easily find them again if they wished. Clearly indicating an 'implant', which should have been searched for?

The question must arise, what if the alleged aliens do exist and really are abducting people? How many of us could accept the possibility? The numbers of alleged abductees have been steadily rising. Can anyone believe it? Any possible alien entities would carry out their abductions, where possible, under cover of darkness and would be well aware that their victims would be assumed to have dreamt it all. Furthermore, in most cases, a form of 'mind control' is said to have taken place, with the aliens attempting to blank out their activities from the victim's mind? There have however, been numerous daylight abductions and the actual numbers worldwide are staggering.

How interested are Government Authorities in getting to the heart of the matter? How many of the alleged victims have had visitations afterwards of the so-called 'men in black', or men in any colour? The Hills never mentioned any such visits.

How would the average person ever get to know anyway? If it is true that such visitations are always, accompanied by veiled threats making it clear to the ones being interviewed that to speak of it would be 'unwise'? Some UFO observers have actually stated that such threats were made not only to them but could involve their families. If there is the slightest possibility of Government bodies suspecting there would be some truth in the abduction cases, how would they react? How could they react? There would have to be at least some dialogue on the topic, however tongue in cheek or light-hearted a manner the subject was treated during such meetings. The question must arise, what, if any, is the ultimate purpose of such alleged alien activity? Clearly, with the huge number of cases on record, some kind of purposeful programme is underway.

Is it significant that so many appear to be of the female gender? Is it genetic material, perhaps female eggs, they require? Female abduction cases always seem to contain events where some form of invasive methods into their person are included. In at least one case, an alleged victim recalled observing foetuses in clear blue liquid in some kind of retort or glass cylinder. Are they attempting to create quasi-human entities, perhaps appearing like humans but also inheriting the alien's advanced mental capacity with telepathic power and other useful advantages? Such entities could be programmed and tutored to know, through simple analysis of the alien memory banks, all the data of Earth gathered since their initial arrival in our system.

They would be suitably equipped to be inserted into the earthly environment with a pre-programmed mission to eventually carry out.

As the alleged abductions have been occurring for, at the very least, 60 years, is it possible that such created entities move among us now, their mission being to gain, by a mixture of natural intellect and mind control, the highest positions possible in all fields of human activity?

If their mission was to influence human behaviour patterns then reaching the highest positions in the military, as politicians and certainly branches of the medical profession such as neurosurgeons, genetic and psychiatry would be necessary.

If the purpose is to create entities for insertion into human life, the implications are extremely profound. They would be like 'sleeper agents', a method we ourselves have employed, to be ready for the call, perhaps for espionage purposes when the right time arrived.

Any 'sleeper' alien creations would not be very likely to be intended for so crude a purpose, but if they could attain positions of power and influence, their real purpose would most likely be to smooth the way, by whatever means necessary, for an en-masse arrival or revelation to us of an alien race in Earth space, and to attempt mentally to convince those in high office or the military to concentrate what we could gain rather than lose by futile aggression, but they would have a mammoth task ahead of them.

Woe betide any human that stumbled on to such a plot. They would be in a worse predicament than an alleged abductee, and would most certainly be assumed to be mentally defective on revealing such a thing. At the least they would be written off as a crank, at worst they would have committal proceedings (for their own good) instigated against them. They would be ostracised, laughed at behind their backs and generally derided.

As mentioned previously how would one react, if one saw clear evidence that a person in a high position was not all he seemed to be? One could conjure up an imaginative scenario where a 'hybrid' human, thinking he was unobserved, summoned up a holographic image of his control (who appears unmistakably alien) and was seemingly communicating with it in a telepathic manner, perhaps using a handset with strange symbols on it and possibly relating information from highly classified documents, perhaps relating to defence emplacements.

What would the observer do? Would he burst in on the assumed 'hybrid' and try to arrest him? Any alien artefact would probably have a built-in self-destruct facility, and once that flimsy evidence had disappeared, the hybrid would simply contact security and have the observer arrested for attacking his superior, and regarding the classified material, the alleged 'hybrid' would have access to it in any case, most likely, having the highest security clearance.

Would the observer be a bit cooler than just bursting in on the suspect? It would be extremely unlikely for him to have a camcorder handy. Would he go straight to the highest authority in the establishment with his story? Even though appearing very convincing the person reporting such a thing would be immediately suspected. However, a certain patronising attitude may prevail, and the person alleged 'to be an alien' would not be summoned to account in front of the accuser. That would not be 'good form', but he would be visited by the high authority and his cohorts (minus the accuser), so that they could all speak freely, one could imagine the conversation, "Look here John, Wilkins seems to think you're an alien. I've given him paid leave and told him to rest and to be sure we will look into it". Chuckles all round. "Poor chap, he's been under a lot of stress you know". More laughter.

Many alleged victims of human abductions feel that they have been implanted with a kind of device that enables the abductors to find them again and, in some cases, the victims are 'informed', although not actually hearing or recalling verbal speech, that they (the aliens) always know where to find them if they require their services again. Any supposed plan or ultimate purpose, the supposed aliens may have, with regard to human beings or the Earth itself, would be formulated over a long period of time and would be as a result of long and careful analysis of Earth and of human behaviour patterns and activity, as said, it would be doubtful if 'they' do exist, that they are infinitely patient and, to be sure, in spite of our lately achieved cosmic awareness, panic, fear and hostility would all manifest themselves in our reaction to alien entities finally deciding to make themselves known to us, and most likely the best option would be to insert themselves in orbit and just wait until we calm down and accept the situation. Is it possible that their ultimate aim will be to 'breed out' our more savage instincts by encouraging genetic 'manipulation' of all future human offspring, to eliminate savage genes and enhance the positive and unused human brain material? Perhaps all the necessary 'hybrids' are already in place as neurosurgeons or genetic 'manipulators' in all the medical institutions and are simply awaiting 'the call' and since a gene has already been isolated governing intellect they may already have also identified those causing negative behaviour but not publicly documented the data.

However, we may not wish to have our brains manipulated at all and may not take too kindly at all to the idea of their 'help'. It would obviously be automatically assumed that 'they' wish to gain control of our minds in order to subdue us.

Of course, their motives maybe entirely different. For instance, they may wish to subdue the world but without bringing harm to humankind with bloodshed or violence, and they may assume, possibly correctly, that once we see what they are capable of, we would quietly go along with their wishes. Is it possible that the plan is already under way? It is said that the male virility count is seen to be dropping, or diminishing. Is it possible that the diminishing ozone layer is part of the plan, perhaps a reconditioning of our atmosphere to suite themselves? Perhaps if these and other factors are seen to be occurring, such as a sudden increase in child prodigies around the world, the alarm bells should start ringing.

However, what could we do about it in any case? We could not charge about the globe like some crazed King Herod, eliminating all the suspected offspring. To be sure, by the time we did finally realise what had been happening it would be too late anyway.

What then, of the poor unfortunate being that may stumble upon, or suspects any such plan. More than one institution around the world may probably have someone within its walls who is there because of such a situation. In most cases, they probably are deranged, but the possibility remains there may be just one that is not. Even so, there is a saying, 'When ten men tell you you're drunk, you'd better lie down'. Eventually such a person would themselves begin to accept that perhaps they were wrong and just possibly were a little bit 'round the bed'. It would all depend on the victim's strength of character whether or not to continue the fight, up until someone took some notice, up to the point of committal. However, once committed, for them the fight would be over, and the slow process of believing themselves being the reason for their being there, would begin to establish itself in their mind.

Another Earth creature, or to be more precise, sea creature, that could be said to be somewhat 'over-endowed' for its needs to survive in its particular environment, is the dolphin, and perhaps for this reason seeks out the company of humans. They are extremely intelligent and receptive to training. Dolphins have a huge range of communicative sounds and certain secret experiments carried out with the ulterior motive of exploiting their intelligence for sea warfare usage, have almost certainly been carried out. Quite some time ago, a well-known electronics firm developed a device to analyse the dolphin communication process, with its many whistles, clicks, barks, groans and grunts. The human proclivity in the military mind always prevails, if it can be used for military advantage, then utilise it.

An equally well-known aerospace company introduced a project to learn 'dolphinese', and the process involved teaching them human sounds as well. This was achieved by the 'reward' system, whereby the dolphin was rewarded with a fresh fish when it got the human sound right. Of course, other creatures are also beginning to be recognised as having quite refined and developed communicative systems, such as gorilla, for instance. Ants and termites also, when functioning as a large group to achieve certain objectives, could not function in an organised and coherent manner, without clear 'commands' or communication. Anyone in a managerial capacity, or who has had military experience and commanded a large group of men, knows that fact. Without precise directives the small creatures would just blunder about getting in each other's way, they must follow the plan, their purpose for existence there must always be those who command and those who obey in an organised society, otherwise anarchy and chaos would prevail.

However, with regard to the use and attempted control of special powers, it is the human, as one would expect that excels. In the USSR, they seem to have led the way and to be more aware of the possible potential of people with special powers, and physic experiments behind the iron curtain were being written about over many years ago. Although the 'iron curtain' has been somewhat drawn these days, physic experiments have been going on for some time there, probably quite some time before the West began to take them seriously.

Duke University in the USA, was among the first to conduct serious ESP experiments, and also in Physkokinesis, or so called mind over matter projects, and no doubt also experimented in the little known DOP, or Dermo Optical Perception, where certain gifted individuals could detect colours by touch when blindfolded. In the USSR a remarkable woman, by the name of Rosa Kulshova, from the Ural Mountains, can quite easily detect colours when blindfolded by touch and another Russian woman could accomplish DOP from the age of nine. Her name is Liza Bleznova, and as well as being able to detect colours with her fingertips, she could also, quite amazingly, accomplish this by the use of her toes.

There are many centres in the USSR studying thought transference and NASA is also interested in its potential for its uses in the space environment. The interesting question is whether humans controlled all these special powers long ago as cosmic refugees and lost them in degeneration processes, and are now in a relearning mode, or that the unused areas of our brain, that is, all the over-endowment of so far unused material, is slowly but inexorably coming into use and that child prodigies, Newtons, Rutherford and Einstein's and all the aforementioned gifted individuals are momentary incursions into those so far unused neurological zones.

The uses of thought transference in space, or Earth space, is not new and was proposed over thirty years ago by Dr Andrija Puharich. He felt that the phenomena could be suffering from the adverse effects of gravity and suggested that, if a laboratory could be set up at the Lagrange point between Earth and Moon, in other words the position where gravitational effects were equal, better results could be expected from thought transference projects. In reality, it is probably quite difficult enough for universities' experimental centres and even NASA, to obtain the necessary funding for projects on Earth that may be cynically regarded by Congress, in the same way as would 'alien busters inc' let alone obtaining funding for experimental establishments orbiting at Lagrange points. In any case, it would have little potential if only tied to one particular position.

Any special powers that in certain individuals may now be manifesting themselves, may have all been mastered long ago by Intelligences 'off Earth', and they may have long left verbal communications behind them. With regard to the general description of alien entities from the hypnotic regression sessions with alleged abductees, a general pattern seems to be emerging and it is a far cry from the rather silly descriptions of the 1950' 'Venusians' with their long golden hair and blue eyes.

If we consider that the massive use of labour saving devices and work droids would reduce the muscular capacity and gradually produce thin 'wasted' body forms, assisted by better diet, fat free and vitamin pills and liquid intake for protein, etc., and that perhaps increased cranial capacity through highly developed brains would enlarge the skull, all that would be needed would be for the aliens to have originated on a gloomier world with a far less bright star, and their eyes may well become enlarged, rather like some of our Earthly nocturnal creatures, we now seem to have arrived at the general description given by many 'victims' of alleged abductions, that is the so called 'greys'.

The thin slit for a mouth that barely seemed to move during communication, more sensed than heard, and the small mouth and tapering chin could be due to the fact that hard ridges only exist inside a mouth where teeth have long been discarded, hence another skull modification in the smaller mouth and jaw. Long into the future we may well fit just this description ourselves, and it is in this line of thinking that gave rise to the theory that these beings are ourselves having mastered time travel and returned, if so, we can rejoice as obviously we have survived that far as a species.

Also, long into the future, if we have to achieve it on our own, without any forthcoming alien 'assistance', when we fully utilise our brain cells' full capacity, we may then realise the true intelligence of other species, the super hearing of dogs, for example – their senses seem more finely tuned than ours to what we might call a 'presence'. They are sensitive to human moods, whereas we have not the vaguest notion how they are feeling.

Anyone who has ever owned a dog will know that if we get up with only the making of a cup of tea in mind, only a raised eyebrow will be seen from the dog. However, if we arise with the thought "I suppose I'll have to take this hound across the common", the dog is up and bounding about before we have reached for our shoes, never mind the lead. Somehow they know and this sense has manifested itself in many UFO encounters, and it was the initial excited reactions of the animal whether dog, cat, horse or cow that drew the human senses towards the occurrences. In an area south of Niagara Falls in Northern USA, a boy doing the milking chores saw a bull in the field acting in a stressed and panicky manner, and it began to bend the iron bar it was attached to from a rope through a ring in its nose. The boy's attention was then directed to a 'football' shaped object moving over some trees about 500 feet away. When he stopped what he was doing and ran towards it, the object shot straight up into the clouds. Elephants have been noted to start pulling their tether chain before the onset of an Earth tremor.

Clearly, it would appear that some animals, though appearing to have little intelligence, are quite obviously endowed with senses we would appreciate as being quite useful. To return to canine manifestations, it is always the dog that is first to hear any approaching party or possible prowler, long before our human ears or senses detect them, and, of course, we do utilise the other acute senses of animals, such as 'sniffer' dogs in the law and order functions of 'drug busts', and so forth, to locate the evidence. Alien entities may well have developed all these acute senses along with their normal advancement. Our creativity and imagination as humans puts an enormous gulf between humans and other species, but our imagination can be stopped in its tracks by such things as trying to imagine a completely new colour other than the well-known colours making up a rainbow, or try to imagine eternity for example. White light we see from the sky is not white at all, but is made up of all the colours in the spectrum broken down by a prism, or in the rainbow where water droplets act as the prism, but we could not add an imaginary colour to it. Science tells us the universe is like an enormous spheroid, but this is still shape. What is outside this shapes boundary? Even if it is just nothingness with no atoms, hydrogen stars, galaxies, boulders, dust, it must still be penetrable. Where does it end – can it end?

Chapter 6

THE SECOND COMING

It seems incredible enough to even imagine that alien entities are in Earth space, let alone the possibility that they may have had several generations of their kind involved in the process of observation and analysis of Earth's species.

Even if their lifespan were two or three times as long as ours, it would have been their predecessors that made the initial trip to Earth. All this is assuming, of course, that what the biblical characters were trying to describe, and beings mentioned in many Earthly legends, were actual extra-terrestrial beings. Either many different types of 'off Earth' creatures are interested in us, as the many differently described craft observed in UFO encounters give rise to, or it is a long and patient project involving the same entities, the variation in the different shapes of their craft may simply be that they were designed for many different purposes.

If the last assumption is so, there must be a very profound reason for such long and patient analysis and, if they do finally decide to make their presence known to us, it will be a 'second coming' and it just might be that we are approaching a point that may well seem quite an appropriate time, if they are aware of the Earthly calendar system and historic quotations and prophesies of the past. If as said, they chose to simply enter Earthly geo-stationary orbits over the post powerful nations and repelled every missile attack laser beams and everything else we could throw at them eventually we would throw up our hands and say "You win" (for now).

Is it time to draw out our file again, on 'ET Encounters/Actions and Options, blow the dust off it and perhaps review it? How much more prepared are we since the last time we reviewed it? Have all those proposed, but expensive modifications to jacket and insulate our reaction capabilities been implemented to defend against possible electronic neutralising techniques?

If the passing 'UFO' can stop all those car engines so easily by just passing, or knock out an entire city's power supply, as has been suggested as happening in the past, what would they be capable of if they actually directed that power toward our electronically relied on, defence capabilities? In my book *Pillars of Fire* I dealt with the substantial amount of events, in Biblical times that seemed so extra-terrestrial in nature involving beings accepted as 'Angels' that had high technology even then.

Have we proposed or introduced any special jamming or electronic umbrella techniques? Have any specially unobtrusive ground to air missile silos been introduced that so far have seen no manifestations of apparent alien interest? It would seem that plenty of UFO interest has occurred at certain defence establishments, air base, nuclear power stations and natural water supplies, and so forth, indicating that 'they' would know all there is to know about Earth's defence and retaliatory capability, not to mention all our vital life supporting utilities. In short if the ETs are intent on confrontation, we would not have a chance, so we can only hope that, as I believe they are entirely peaceful the ufologists have long suspected, there has been a massive cover up operation in progress and small humanoid bodies do float in special tanks with actual UFO hardware now being examined but we are no nearer at the moment in our understanding of what UFOs actually are and what their mission is, than we were in 1947 when they 'raised their profile'. Although it seems obvious that UFO activity was going on long before this point in time, it was in 1947 that we first heard the term 'flying saucer'. It had a nice ring to it and certainly fired the imagination of the science fiction film makers, who in the fifties produced a great multitude of such material, and of course, after that, everyone was seeing them which many 'de-bunkers' of course, took the view that it was merely auto-suggestion.

Throughout this work I have purposely tried to use the term UFO, which means unidentified and not necessarily extra-terrestrial, but whereas the manifestation of objects in the sky could be natural but unexplained phenomena within the realm of physics, alleged abductions and close encounters of the third kin are a different matter and can only be lies, figments of over-active imaginations or true occurrences.

If we take the occurrence in France, where a lavender grower saw what he thought was some children picking his crop, moved closer and small entities ran to a circular craft, boarded it and the craft rose speedily in a vertical mode until it was out of sight, that farmer would neatly fit into one of the said three categories.

Many people will take a sceptical view and write the whole UFO phenomena off as being too silly for words, perhaps not realising that they are writing off the sworn testimonies of whole armies of people and forgetting that the truth is often stranger than fiction.

We watch films about super spies and special gadgetry employed, and sum them up as entertaining but far-fetched, perhaps forgetting real life occurrence of a chap from an Eastern Bloc country, who felt a jab in his heel then received an apology from a 'city gent' type with an umbrella looking like a thousand other city gents. Later the chap dies in hospital and the post mortem reveals a small capsule that was embedded in the victim's heel that dissolved with body heat releasing deadly poison. Clearly many 'Q' sections exist, busily manufacturing small, but deadly gadgetry for use by clandestine organisations and their operatives.

As said at the beginning of this work, we can plan and equip for operations against a known or potential enemy, but how does one prepare for the possible adversary about which we know absolutely nothing? If there is one trump card we do possess, it may be in the form of bacteriological weapons, but utilising them in attempts to defend our planet may result in wiping ourselves out. However, is it possible that this would be the one thing they may fear above all else. It must be said that if, as many believe, the alleged ETs where responsible for our very existence due to genetic manipulation it is not logical to assume they wish us harm. The only worrying factor is that during the initial experiments they had hoped for better results but had to eliminate many failures in widespread human elimination even to rid the world of undesirable people who may beget more malfunctioning humans and we have to admit that (perhaps in a lesser volume) many are around today.

However, the human spirit will prevail; there would be clandestine meetings; the 'right' men for the job would emerge as is usual in times of crisis; the Churchill's, Eisenhower's, Montgomery's and Roosevelt's, would all rally to the call and the Earth Resistance Movement would be immediately instigated. A select band, who may not wish to have their heads shrunk at all, would arise and vow to fight them on the beaches, etc., etc. Whereas in times of war, these qualities are admirable, against an enemy with evil intent there is not a particle or evidence to suggest that ETs fit this description.

Nevertheless, select bands would retreat to the mountains to plan their campaign, and all those underground complexes inside the tunnel systems of hydroelectric schemes intended for defence planning in the old cold war days, would have a new purpose, and more frighteningly the germ warfare plans would be hastily reviewed, Porton Down would be very active.

For all the supposed superior intellect of alleged alien intelligence possibly in Earth space, they may misjudge their timing with regard to any 'help' they plan for us. We may yet have to swirl further down into the maelstrom of drugs, organised crime, degeneration of moral standards, lack of parental control, beating and mugging of the elderly, frail and vulnerable for a few pounds, child prostitution, disease, rampant Aids, and so forth, before we are ready to throw up our hands and submit to any external 'help' from an extra-terrestrial source, when it is finally shown to us that our noble-minded human spirit and intentions to 'fight to the last' are just a manifestation of a wish to hang on as long as possible to our outmoded, negative and destructive behaviour patterns, this may just about bring us to the point of submitting to their will at last. However, we would never give up without fierce resistance having possibly noted any vulnerability the ETs may have, but finally we would be subdued.

However, before that time came, perhaps great damage could be inflicted upon the imaginary alien beings. Although in the past, aircraft have been scrambled in an attack mode to pursue unidentified 'blips' on radar screens, we rarely hear of any reported attempts at some kind of communication with an unidentified craft, and in the aforementioned imaginary scenario, the 'human spirit' once united, would be a force to be reckoned with, especially with a common purpose against a now revealed 'foe'. However, in the final analysis, it would have become clear that in terms of loss of equipment and lives that the whole exercise was futile and we should have listened to their message in the first place but as we have stated that is not the human way.

Clearly, they would be aware of our weapon and possible bacteriological stockpiles and, whereas they would have long developed antidotes for all airborne traces of atmospheric bacteria, there may be some they would not have had the chance to properly analyse. Another slightly worrying factor, even if their attentions were strictly honourable is that if they did arrive in their 'second coming', carrying bacteria as we do in our body form, food and equipment that does not harm us the same would apply to them and whereas they would clearly have taken great pains to protect themselves from our bacteria, could they be (unknown to themselves) carrying an older form of 'alien' bacteria they had long become immune to, but which could be absolutely deadly to us, and though they come in peace they wipe us out anyway. However, if they are as intelligent and advanced as we surmise they surely would have considered that. In any case, with all the alleged abductions someone would have succumbed to it by now.

If they do have advanced intellect, which is more than probable, they may think in terms of pure logic and common sense (to their way of thinking), and not be swayed by any emotional or ethical factors. If they think we have now reached the stage of emotional stability and intellectual capacity to comprehend their logic without question, they may expect us to readily agree to neurological treatment and stimulation when showing us the benefits. They may have compiled explicit scenes in a manner we would be familiar with, such as a simple DVD film of human atrocities since the days of pagan Rome, down through Genghis Kahn, Attila the Hun, the Viking hordes, the horrors of the First and Second world Wars, gassings, mass killings, the holocaust, atomic weapons (used in anger) to stop the horrors in the Pacific. They may oblige a 'captive' audience through taking over the world media after neutralising all our deterrents, to watch such a compilation, which may finish with our interest in

Star Wars weaponry and grandiose plans for space travel, with wars still going on somewhere on the globe. The whole idea of the exercise being to make us feel thoroughly ashamed of ourselves.

Finally, we may just get the point and be well on the way toward agreeing with their proposed methods of 'adjusting' our cranial cellular material. In return, there would be promises of eliminating our poverty, starvation, wars, crime, disease and other worldly problems, practically overnight.

Dr Vyacheslav Zaitzev, a philologist, late of The Byelorussian Academy of Sciences, was a supporter of the 'Jesus was an astronaut 'theory, and stated that the Earth is on the eve of 'The Second Coming'. Nevertheless, it may not be in the manner that he surmises. We have not ruled out the possibility of a biblical 'second coming' but are we to imagine all the evil doers on Earth will simply fall down dead.

The Persian legends of 'flying carpets', the Indian 'Vimanas, and all those flying shields may all have an extra-terrestrial connection, and may be the closest the ancients could get to describe what they were seeing by relating them to things that they understood. Is then, the battle for Earth imminent? Is the 'final conflict' almost upon us, when the struggle for domination of the Earth begins, or it is just their 'Second Coming', when the hypothesised aliens, who have overseen us for so long finally, perhaps in sheer exasperation, eventually make their move? One wonders how they would actually go about it. They would know that in spite of all their appearances and obvious clues they had given us, that there would be absolute panic and pandemonium below when, their probably hugely impressive interstellar mother craft, came into orbit, and they would wait for however long it took for us all the calm down a bit and maybe even return from the hills, especially if everything we had thrown at them has failed even to dent their craft. Dialogue would begin with the world leaders and they would be relied upon to make the best decisions, but there would be panic, retaliatory attempts, but nothing with electronic reliance would function.

That would have been foreseen and our behaviour patterns would be well known to them. Every facility would be immobilised. After we had tried and failed with every method of attack at our disposal and began finally to think about instigating some form of dialogue, as all aggressive methods had failed, we would possibly get no reply – we had our chances at that and blew it. However, it would obviously be an 'occupation', not of a country but of the entire world. Armies of occupation on Earth during our ceaseless conquests, have always given rise to resistance groups, France, Poland, Holland, Greece, all the lands occupied by the Germans saw such activity and could never be fully eradicated, even though the enemy savagely took and murdered many male hostages and burned entire villages to the ground. So it would be, although far more traumatic and difficult to achieve meaningful results, with regard to an alien presence, no matter how honourable their intentions were and no amount of convincing rhetoric regarding the benefit to humanity that had been made it would not prevent a permanent will to resist by any means possible. A resistance group or groups would arise, that we could term the E.R.M. or Earth Resistance Movement. It's immediate priority, would be to try and discover a 'chink' in their armour, some kind of weak point would be searched for and probably found, possibly of a bacterial nature. However, this may be a step too far with the possible results and clear and present danger to ourselves. Nevertheless, UFO activity has been observed over nuclear power stations and weapons establishments, air bases and suspected stockpiles and missile silos. However, maybe, although governments refute it, biological agents and means of delivering them are in existence, simply because we have never trusted possible enemies (even though they also deny it) to be equally equipped. How many UFOs have been spotted hovering over Porton Down? For all that, we have to consider the possibility of the aforementioned 'Hybrid' insertion programme. For all the decades since abduction claims began, such beings would have been in existence for lengthy periods of time and would most certainly exist in the scientific institutions, all the branches of the armed forces, probably in high ranking positions and also in all the political parties, they could infiltrate an E.R.M. with ease. The possibility must have been discussed and the appropriate files raised. If all those 'Men in Black' visitations were government agents, much data would be filed away. However, none of this would affect the human determination to resist in one form or another and this could be our downfall, we may become our own worst enemy.

We have spoken of the longevity of the possible existence of ETs in Earth space ever since the 'Angels of Abraham' and indeed much further back in the time when we study the period of Noah and the flood. All of course, pointing to the distinct possibility of these entities being responsible for our very existence and elimination when they considered it necessary in those times, which of course they would still be capable of. All of this keeps returning to the unavoidable conclusion that there must be an 'end game' and final decision, an 'Armageddon' a 'second coming'. A decision regarding the latter was made nearly 2,000 years ago during the time of the unique being we know as Jesus. Not by Him we must add, Jesus was all for it, the decision was made in 'Planet Heaven'. Jesus made it clear that, when being continually prompted by the disciples, that "No man Knoweth of the time, only my Father in Heaven". Nevertheless, he did state, when speaking of the second coming "This generation shall not pass until all these things be fulfilled". Later, making him seem a false prophet. Jesus survived barbaric execution and was revived by distinctive beings that rolled back the stone of the sepulchre administered the revival procedure, then sat back on a rock awaiting the arrival of people who would prepare Jesus for burial. It happened to be Mary Magdalene. The beings informed her Jesus had arisen. It must have been quite recent as Mary caught up with Jesus but He warned her "Touch me not", clearly indicating that the revival process would be dangerous to humans coming too close. Before he 'rose up' where a 'cloud' received him out of their sight the crowd had been assured of a 'second coming'. The scriptures state that two men in white appeared, spoke to the crowd regarding the return of their 'Messiah'. Their type of clothing must have been distinctive or it would not have been mentioned. They stated "Ye men of Galilee , why stand ye gazing at this man's departure when he will return in like manner as ye have seen him go". These beings and (as said) Jesus himself, were not aware that the decision 'postponed' had already been made in planet heaven.

One significant statement had been made during discussions of the return of Jesus and that was that there would be 'signs in the sky'. There is no denying that we have had a multitude of those in our time. Another important point would be, that an alleged female abductee having been suitably calmed and relaxed began to ask questions of the beings and why they were doing it. The reply was as we mentioned before that their operations would be soon drawing to a close.

Only one 'prophet' or clairvoyant comes to mind that was specific about a form of 'second coming' and that is Michael Nostradamus. He was specific in mentioning the year and even the month of the event. His other predictions where strikingly accurate which has to be a little worrisome, but this 'Armageddon' type prophecy is now two decades overdue. He stated, "In the year 1999 and seven months a great 'King' shall come from the 'sky' and Mars (war) will reign supreme". Is the 'war' the probably futile resistance already mentioned? His reference to a 'King' most probably only meant a being of great power rather than a crowned entity in ermine with an orb and sceptre. We have to consider, that since the other prophecies made by Nostradamus where so accurate, that perhaps the unknown power that had bestowed him with such gifts, had failed to mention that another postponement had been made. Nevertheless, his predicted time for the great event is still very recent and the older prophecy that "There will be signs in the sky", has certainly come to pass indicating perhaps that the 'time' is nigh.

In this regard, we could assume that they would have three choices. Firstly, to calculate that there would be fierce resistance and not wishing to harm humans (or they could have done it long ago) continue in the observation mode and continue patiently inserting hybrid entities. Secondly, to go ahead with their 'Second Coming' on the basis that human behaviour patterns may never change and the only way to cure the 'human question' would be to neurologically adjust the genetic code in all humans first born, but for that complete subservience of humankind to their will would have to be achieved and thirdly, simply go away. If they are advanced enough to achieve superlight speed or traverse great distances utilising wormholes etc., great time periods could pass on Earth but not for them. This would allow for the propagation of the substantial amount of hybrids they may have inserted into human life to bring their influence to bear in the human stock and also the excessive amount of additional brain material in the human that was introduced during their hypothetical genetic creation programme, carried out by their forebears long ago, may, as originally intended have become further activated causing humans to finally perceive the futile stupidity of war and conflict. After all, we may not be 'the only game in town'. Their entire race would not have spent all that time in Earth space. They could be involved in many other planetary operations of terraforming, creating and perhaps helping emerging life forms toward mental advancement far from Earth. They may then return to Earth and completely reassess the situation with regard to humanity and we would still be seen as having abdicated from all human rational behaviour by continuing with absurd war and conflict then the same dilemma would arise for them.

The additional brain material in the human may well have become active in the positive mode with great advancements in science and all the persuasions and even in space activity with terraforming operations active on Mars and Venus, but not having shown any great depletion in human negative behaviour patterns with other nations aggressively competing for the mineral wealth discovered on the Moon, Venus and Mars.

In this regard, they may consider (not being infinitely patient) that their second option to go full steam ahead with their second coming and deal with any situation regarding human reaction that arose, only this time their star ships would not only descend over all the capitol cities of the world but also all the colonies established on the Moon, Venus and Mars.

Humans involved in the (long disposed of) S. E.T.I programmes, may have arrogantly decided that humans may be the 'first' and act accordingly, since no definite contact had been made with another intelligent race on another world. In any case they would quite soon have the power sources to go and look for themselves having obtained helium 3 from the Moon, gained control of anti-matter and developed fusion power sources. This most certainly would heavily influence option two with regard to ETs.

Humans, not having made contact with another planetary race, would be subject (perhaps more so) to cultural shock and social disorientation particularly as they may have somewhat arrogantly decided that humans will be the first cosmic travellers. By this time our hypothetical ETs patience would have entirely run out. the ETs could not escape the inherited responsibility for humanity, since their forebears had created humans in the first place, and although this had happened so long ago the responsibility would still prevail and rest with them. It would be even more difficult for humans at this point to become subservient to a greater power but they would have no choice.

Any kind of shock, given enough time, will subside. Two major shocks occurred that greatly affected the ecclesiastical fraternity and millions of people who considered themselves quite religious, not only in this county but all around the world. The first, although traumatic for many people, was perhaps not as fierce as the second in terms of people's religious viewpoint. Some 160 years ago when Charles Darwin's 'on origins of the species' appeared, many people were outraged that he could suggest than man's predecessor was an ape and that gorillas and chimpanzees where man's closest living relative but at least they could find solace in the fact that it was only a theory and not at all proven fact also, that Darwin himself was not 'over pedantic' about it and made it clear that the evidence, in terms of the vital fossils showing clear transitional stages would have to be found. This took the pressure off himself but transferred it to those who considered themselves professional in the analysis and classification of old bones who had to begin the quest to prove it all. Many, welcomed the theory, having little or no religion anyway, and it seemed an acceptable theory rather than blindly accepting the mind stretching pronouncements of Genesis. Therefore, the religious people who had considered themselves God fearing could still look forward to a hereafter if they behaved themselves in accordance with the Ten Commandments, as the Darwinians hypothesis was not considered strong enough to divert them.

Of course this situation prevailed over the decades, amounting to about a century, before a newly offered and rather startling theory emerged that became a third alternative for people to consider and many embraced it, simply because they had little or no religion and the Darwinian theory had not convinced them either. Historical and biblical writings where shown rather convincingly to have been simply misunderstandings as to their significance which intrigued many people, as strong circumstantial evidence seemed to exist that there was indeed a higher power than man and that human creation did actually take place but far removed from notion spelt out in ecclesiastic terms, that is , Genesis.

However, we always have to return to the consideration that there must be a final conclusion to it all concerning any of the three options we may accept as to how humans 'came to be'. However, before that, we must return to the aforementioned cultural shocks, which, as said, was firstly the Darwinian theory that religious people could bear due to the unproven aspect of it, but the second 'shock' was much more fierce in its effects on religious beliefs. This took place some fifty plus years after the introduction of the Darwinian concept, when many people who had survived that disruption where still alive. I refer to the infamous Piltdown hoax, this concerned what appeared to be the final convincing evidence to prove the Darwinian concept. In 1912 at Piltdown in Sussex, a human almost complete skull, together with an ape's jaw where found and accepted as genuine fossils. Many people had no choice but to accept that they had to abandon their religious beliefs. This outrage caused many people to go to their deaths over a forty-year period that had turned away from the religion until it was discovered as a hoax by more advanced scientific dating processes, in the fifties. Such was the desperation to prove the Darwinian concept, the bones had been purposely buried, knowing they would be found, but the 'third alternative' that was offered did not shock or disturb religious people as they viewed it as an interesting science fiction like data that one could either accept or reject.

However, we must return to the main point of this work and that is the final culmination of it all.

The Freedom of Information Act proved that governments had been studying the UFO phenomena and the ET hypothesis and still where. They must also, have, undoubtedly, considered that there must eventually be an end game or culmination to it all.

Most certainly Hollywood has, by producing interesting movies dealing with this eventuality. Three such films come to mind. The modern take on H G Wells' *War of the Worlds*, with copious amount of blood spraying everywhere, and of course, *Independence Day*, both display aliens with evil intent and compiled largely for their violent action scenes and how the heroes would finally subdue them. The third film however, depicted a totally opposite view, peaceful smiling benign aliens, finally appearing to a select few who 'they' had singled out to be present at a location selected by the ETs themselves and transmitted by coded signals as map references. This film was compiled utilising many factual cases and events regarding the UFO phenomena. It consulted for assistance the scientist/astronomer and one time sceptic and debunker of the UFO phenomena for 'projects blue book'. In his case, the multitude of UFO events he had dealt with made him quite convinced of the reality of the ET hypothesis, he was Dr J Allen Hyneck who appeared in a cameo role in the final scenes of *Close Encounter of the Third Kind*. He had classified the mass of reports into encounters of the first, second and third kinds. The film also introduced the theory of 'time dilation' (in visible terms) where the flight crews of the lost 'Flight 19' in the forties, much dealt with in conspiracy theories regarding UFOs, appeared out of the landed craft returning as young as they were when they disappeared, that is, being younger than their descendants. This has been offered to explain the longevity of the UFO phenomena. The Einstein concept has it that a craft moving away at the speed of light would effectively be travelling back in time and if it later returned, its occupants would not have aged very much at all but people of the Earth would have aged in the normal manner. Most of Einstein's theories have been borne out but he set a limitation of light being the ultimate speed as the infinite increase in speed would result in infinite mass, as mass increases with speed, but certain experiments at C.E.R.N. Europe seems to challenge this 'maximum' theory.

The landed craft in the film also disgorged numerous crews of vessels that had disappeared in the notorious *Bermuda Triangle*.   In fact, the UFO had dropped a ship in the middle of the desert and the missing aircraft of Flight 19 in a South American junkyard in the same pristine condition they were in when they disappeared. All good entertainment of course but heavily reinforcing the final end to all the conjecture and wondering how a 'second coming' in whatever form might or would take place.

In general terms, we could agree that the evidence shows a far superior and advanced ET race that could have, if they so wished, have eliminated humanity long ago, indicating their peaceful intent, but there is always a counter argument, which may state that since ET seems to have been on Earth for so long and may be responsible for our very beginnings, than, as possibly misconstrued Biblical texts point out, masses of humanity where ruthlessly wiped out in purposeful annihilation events showing that ET is quite capable of this is viewing their 'creations' i.e. us as faulty or, since they are theorised as being responsible for human current behaviour their mission profile in a possible second coming may well be (partly at least) an elimination event. This is reinforced by our continual display of dangerous negative behaviour that shows clearly that in spite of those alleged biblical annihilations that took place in the days of Noah, Abraham and Moses, the methods, power and abilities of the ET that may soon come among us would have improved over three or four millennia and even longer when we consider the Noah story, but this implies 'Earth Time' and they, when streaking away to involve themselves in other cosmic events far from Earth, may not have been subject to such time periods.

However, the main point is, human behaviour in the negative mode has improved very little in relation to scientific advancement in the other fields, particularly in our cosmic travel plans, and this would surely be a major consideration in their mission profile that would have to be dealt with and as said, this would have caused such a long period of time to pass in as how they would have to deal with it. Surely, the last consideration would be annihilation until every other option had been considered and dealt with. They may have now reached this position. When considering the biblical events and the ET hypothesis maybe one and the same, then today, surely 'they' may have drawn up an extensive list of those most deserving of termination. They would not need any bloody crosses to be painted on the doors of houses in the area of where these alleged evil ones existed, as was necessary in the days of the biblical event of the Passover to eliminate newborns they would be well aware by now who they where and where they lived.

Another similarity between the biblical happenings and current ET belief being one and the same is the event that many religious people firmly believe will come to pass, is the so-called 'rapture' where the righteous will be immediately taken to (planet?) Heaven and the rest dealt with accordingly (but not explained how).

St Paul also spoke clearly of this some 25 years after Jesus' departure. Many factors point to the biblical events, in particular the Old Testament as being one and the same regarding the ET phenomena.

We mentioned the late Dr J Allan Hyneck. One wonders what factors he discovered when personally investigating many of the most high profile UFO cases that caused his very noticeable transition from total sceptic to convinced believer. He wanted science to change its tune and come together in some project to seriously study the topic and its various implications with regard to humanity. Hyneck was Chairman of the Department of Astronomy of North Western University USA. He was also in charge of the optical satellite tracking programmed of the Smithsonian Astrophysical Observatory in Cambridge Massachusetts and Scientific Director of the US Balloon Astronomy Project 'Stargazer' for eighteen years. Clearly, he could not be more qualified to deal with a postulate on things celestial.

It is said, that it is no longer fashionable to be sceptical on the subject, yet there are many scientists who will not go near it. However, there are always exceptions, a leading scientist of a top ranking University who had closely reviewed his data on the UFO phenomena and its possible conclusion, berated Hyneck for not coming out boldly with a statement stating clearly that UFOs had to be extra terrestrial. We could cynically ask "Well why didn't he do so?". After all, he was the type of scientist that Hyneck was trying to attract to his banner and form a positive group to do just that.

Science has often postulated that people who claim to see and report such things as UFOs (and their occupants), represents nothing but a psychological problem. Others were somewhat shocked by this and stated (regarding the huge amount of people that had reported) "If the phenomena is purely psychological, then this should be considered even more startling than even that which suggests visits by extra-terrestrial.

This is a valid point, as it would be a disturbingly large portion of the US that would have to be mentally unreliable which would involve all the professions and the top military and a couple of US Presidents. At a quiet scientific meeting of the Optical Society of America on the subject of UFOs by Dr Hyneck, had offered far-reaching implications as far as the scientific attitude toward the phenomena was concerned.

He demanded that the UFO phenomena requires serious scientific attention. He stated, "I say this at the start, so that you are not misled by the kooks and nuts and the gullible, which have made this subject so difficult to explore rationally". "UFOs are a real puzzle. The subject has not been put to rest, the scientific community must now recognise that we can no longer dismiss the subject in the usual ways". However, for all that, scientists like Hyneck still remain in a lonely position among their community.

As mentioned, Dr Hynek spent 18 years as a sceptic further indicating that something quite profound must have caused his conversion.

By now, whether it is admitted to or not, an intensive study by selected scientists has almost certainly been undertaken, who would be protected by anonymity so that the mainstream scientists would probably not be aware who they where (and may even belong to an institution covering a totally different subject).

Equally certain is, that an in-depth pattern analysis by computer of all the existing data and reports and high profile encounters has occurred and they would have considered and discussed a 'final appearance', possibly and whether the general population should be gently informed of such an event happening. This would lessen the trauma of the actual appearance and give the population a chance to decide in their own minds how they would (mentally) handle it. The best approach would be to firmly state "Look, we know ET and their descendents have been with us since biblical times, so it is very unlikely that any harm will come to you, otherwise they would have done it by now, so act naturally".

Whether that approach would work or not remains to be seen, Hynek was known to have stated when a phenomena has the potential capacity of a possible scientific breakthrough, we are neglecting our own responsibilities by not exploring every facet of the phenomena. Today however, among the many and varied conspiring theories that exist, we have already learned much from their alleged downed craft and live body forms analysed. Anything goes in the conspiracy theories they know no bounds. If we could succeed in the trauma and cultural shock of a far superior intelligence coming among us, which on past human behaviour, where, during the age of discovery, a more advanced peoples imposed themselves on a lesser advanced it was always detriment to the latter but in our case we should indeed learn much. The signs are that the special powers once dormant in the brain are developing becoming more active, and that this is due to the additional material in the brain, once referred to as 'junk DNA'. In the many alleged abduction cases, telepathy always is a dominant feature but in only a one-way dialogue, mostly assuring the abductee would not be harmed. This proves that the human can at least receive, but the huge amount of people close to each other who have said something and the other says it first, must tell us something.

In any case, it has often been proved that the human brain can transmit. How so? You may ask, well, during the many tests for brainwave function and its activity, which is a form of electronic impulse transmission probes are affixed to the skull and not the brain itself, the signal received by the attachment probes has been transmitted across the space of the gap between the brain itself and the bone structure of the skull and then transmitted to the readout instrument, but we do not know its full range but the signs are it can be transmitted from person to person. It has been suggested in UFO literature, that ET may have mastered this 'thought transference' to such a degree that the skin-like costume or clothing they wear (in a well known abduction case the beings in thereafter wore some kind of cap) their brain waves or thoughts could be transmitted to the power source in the craft, were simply thought power of 'up' 'down' 'across' and so forth is immediately obeyed, explaining the rapid direction change of UFOs. The alleged abduction by ETs must be discussed and would most certainly have been given an important place in any scientific analysis aforementioned, by selected members of a purposely set up group. They would however, have to consider that their worrying over the amount of people affected by an alien encounter, implies a huge psychological problem (if not genuine), it would also have to be applied to those who claim to have been abducted, as they also exist in huge numbers around the world.

One wonders, if an alien plan is afoot to appear to humanity in some form of revelation, whether they would contact their abduction victims to invite them to their forthcoming appearance. Many 'victims' claimed to have been told 'they' will always know where to find them. In addition, of course many a metal insert has been found inside the body as alleged 'implants'. Clearly, there would not be any point in inviting them if they were going to land on some high mountain as in *Close Encounters of the Third Kind*, what amazing irony it would be if Cape Canaveral had been designated and, like in the film it disgorged scores of children now developed that may have been the foetus' shown to some female abductees and in a second abduction, instead of floating in some chamber of blue liquid where bonded with the female abductees, this has been claimed, who sensed they were partly theirs. In this manner, they could insert the future progenitors of humanity who may be as genetically perfect as it was possible to produce and may grow into maturity as supremely intelligent and devoid of any aggressive, violent and savage quantities and never consider war.

Many females who may have gone through the 'bonding' process in a second abduction, would accept and adopt them. Other craft may do the same in other countries as abductees have occurred in all over the world. That of course would be the favourable scenario that would be preferable but not of course guaranteed.

However, there must be some quite profound reason for the decades of alleged abductions that have been going on, seemingly since at least the 1960s. Since that was some 55-60 years ago their programme would have been drawing to a close as was purposely implied to some abduction victims. How much else could they learn about human bodily metabolism after all that time? Obviously, there was a lot more to it than that.

If (as many believe) ETs where responsible for our very existence and gave us all that extra brain cellular material, they would know the complete human genome in all its intricacies.

It may not be an exaggeration to say, that some 'exterior force' such as Alfred Russell Wallace puzzled over regarding the over-endowed human brain, has been found and strongly reinforces the ET human creation theory. This is akin to finding an alien artefact on Earth, only it is a specific gene unique to the human with no origins and certainly not evident in the apes. It is known as the Fox P2 Gene and is ascertained as having been introduced into the human brain fifty thousand years ago and caused the lower positioning of the thorax and the development of articulate communicative speech. Human cultural development increased shortly after (in evolutionary terms). Humans must have existed quite some time before this with modern anatomy and intelligence, but good communication was vital, if humans where to advance as far as they have today with dreams of interstellar travel. I mentioned in other work that a gene has been identified that governs intelligence, this of course, begs to question why hasn't a gene been specifically identified that is equally responsible for negative or criminal behaviour and is such a search now in progress? It took many years (six in fact) to isolate the intelligence gene and how many knew it was in progress? Equally, why haven't ETs with all those abductions identified it either? Well, the answer may be they have and most certainly have been doing something about it in producing all those hybrids from human female egg procurement. However, the Earth's population is vast and ever increasing, as are the numbers of people afflicted with these negative tendencies.

The obvious conclusion is that an intensive programme of genetic manipulation would have to commence as soon as possible. If the onset of the human abduction programme began in the 60s, the most convincing and believable case occurred in 1961 concerning a couple who were driving home at night from a visit across the border to Canada.   They were heading back to New Hampshire through the mountains at night. We have referred to the Betty and Barney Hill case. A complete account of this amazing event is contained in John G Fuller's book *The Interrupted Journey*, Corgi in 1981, which I would fully recommend. Those with computer connections can call up a process where books gone out of print can be obtained.   Repressing their experience caused in Barney Hill's case, health problems regarding a duodenal ulcer and his wife Betty had disturbing dreams and flashbacks and they both tolerated all this from a couple of years until seeking help which came in the form of a Boston neuro psychiatrist Dr Benjamin Simon.   He was very proficient in hypnotic retrogression. Not in the sense of a stage show hypnosis making fools of volunteers he had impeccable credentials, a former lecturer at Harvard and Yale and a Fellow of the Rockefeller Foundation studying neurology. He was or had been a Lt Colonel in the US Army and specialised in dealing with traumatised troops, through regressive hypnosis.

The Hills could not be in better hands, they were not a couple of imaginative teenagers who would not have been listened to anyway and may never have to be in the position of being dealt with by such a professional, but what if they had been a couple of teenagers, how would such a traumatic experience affected the rest of their lives? All the hypnotic regression sessions were recorded on tape and Barney persisted in asking that they specifically retained for comparison should anyone else go through their ordeal so, although Dr Simon May well have passed away by now some trust he nominated would have them. When the Hills where driving home through the White Mountains there was no traffic as winter was approaching and all the visitors centres and motels were closed up. A bright light was observed nearing their car and getting brighter. Barney finally pulled off the road onto a viewing area used in summer times. He left the car and trained a pair of powerful binoculars on the craft which was now lower and nearer and clearly saw its circular shape with lighted windows around it and a figure that was peering down at him he described the being's eyes as "Like a cat's but dark and going slightly around the head". He was desperately trying his hardest to pull the binoculars away from his eyes but could not do so. The eyes where telling him something. Finally, mustering all this strength, he pulled them down so hard he broke the leather strap and caused a red welt to appear on the back of his neck.

He ran back to the car and they both recall hearing a series of beeps, then the next thing they recall was approaching their home a couple of hours later than they should have. The infamous 'missing time', that is often mentioned in abduction cases.

Thus began their period of amnesia regarding the missing time.   When Barney had reached the end of his therapy he and his wife Betty where astounded with regard to what they had both been subjected to. Dr Simon drew out everything from Barney's sub-conscious, regarding what happened from the moment Barney left his car to get a better look. They will not be related here in case any reader wishes to obtain the aforementioned account for themselves in *The Interrupted Journey*.

Any group of ETs suddenly appearing to us on Earth in some culmination previously described, would not be able to summon the Hills as they have both passed on but we can be sure the craft and its occupants that where responsible for their traumatic account are still around.

Although I will not relate the entire session and substantial amount of data that emerged detailing the whole abduction and examination. I would mention that Dr J Allan Hynek became very interested in the case. He met up with the Hills and with the biographer of their hypnotic recall and the extensive seven-month period of analysis that took place, that is, the author John A Fuller. In his books, he added many personnel details about their background, jobs and local community work etc.

The Hills, Fuller and Hynek all got on very well and with the Hills permission and that of their therapist, Dr Benjamin Simon all met up. Dr Hynek had extensive files and data on all aspects of the UFO phenomena, abductions, radar contacts and so forth and wondered if he could corroborate their experience in any helpful way.

The Hills wanted to know more and Dr Simon welcomed the idea of science at last showing an interest in these obscure events rather than simply putting the ball in their court and writing it all off as a psychological problem.

When Dr Simon, using the key words put both Barney and Betty under hypnosis, he told them "You will answer all questions by Dr Hynek as though it was me asking you them". Betty and Barney sat together with Dr Hynek and John Fuller opposite them. Dr Simon was in-between. Dr Hynek was very impressed by Barney's excited recollection referring to 'men on the road' coming towards them and assisting him out of the car and up a ramp into the craft. The beings seemed to have control over his mind and he was helpless. Barney mentioned feeling his feet bump over a sort of bulkhead as he entered and observed a corridor with Betty in front being taken to a separate room.

Interestingly, when Barney and Betty reached home after the start of the amnesia some years before the therapy, he noticed scuffing on the tops of his shoes. He was also concerned about a circular series of warts around his groin. Betty removed her dress put it away and never wore it again.

During Dr Hynek's question Dr Simon interjected and asked Barney "Where are you now? Barney said, "I'm in a corridor, I don't want to go I'm on a table".

The thing that Dr Hynek learned when seeing Barney's traumatic reaction to the questions, left him no doubt that the Hills had experienced a frightening and disturbing event. At one point Barney jumped out of his chair and had to be calmed and re-seated.

When Betty was asked, under hypnosis, about the roadblock and the strange creatures she said, "I want to wake up, there's a man behind me and two others holding me up".

When Dr Hynek asked, "Where are they taking you", she said "In a path in the woods towards the craft". She related lots of detail while being questioned, such as the craft was round and metallic and about 80 feet wide.

Barney when under his questioning said with regard to the size of it, that it was as wide as a B47 from tip to tip (the wings). When he was pressed about the alien communication, he mentioned understanding everything they transmitted mentally but as regarding any sound, he described it as a humming buzzing noise.

All of this was locked out of their minds and repressed for the couple of years before their therapy that was the period of amnesia. Dr Hynek had never before been as impressed by any interview before as he was when listening to the corroborating detail from both Betty and Barney, and during his investigations, he had covered a multitude of cases relative to UFO experiences of all kinds during his lengthy period of dealing with them.

As for Dr Simon his brief was to alleviate their stress and locked up amnesia and not concern himself about the validity of UFOs or their occupants and he became suspicious when Betty said that 'they' inserted a probe into her navel. Dr Simon knew there was no such medical test yet a few years later, there was.

It was announced in Medical Journals that a totally new method had been developed to examine the amniotic fluid, by the insertion of a needle through the abdomen and later to obtain eggs from fertile women a needle through the navel. These where the days when the operation of the so-called test tube babies began.

Another interesting feature of the case arose when Dr Hynek was researching old records regarding UFO sightings during the period of their abduction. He discussed an overlooked Air Force radar report that showed a radar contact with an unidentified flying object in the exact vicinity where the Hills had their experience. There is such a wealth of corroborative evidence (some circumstantial) to back up this case that is hard to ignore and as I have said, if you do not believe this one you will not believe any of them.

Another interesting event during the Hills ordeal was when the ETs found that Barneys teeth came out, that is his dentures. It caused great excitement and quite a buzz (literally) as they went into the room where Betty was being tested and tried to pull her teeth out but Betty did not have dentures.

It is incidences like this that make the case stand out as not a conceived or purposely contrived event. Also when the Hills got home after their ordeal, which was repressed from their memory by purposely induced amnesia, they both agreed that they had heard a series of beeps that seemed to be toward the rear or 'trunk' of their car. Betty went out later to look the car over and she noticed a series of shiny marks on the trunk lid, they were about of a silver dollar. When mentioning them over the phone to a friend she was advised to check them over with a compass, when she did so, the compass needle span erratically, yet later when the enquiry began to be pursued further she completely forgot to mention them to anyone and they were never mentioned in any of the hypnotic analysis sessions, which incidentally lasted over a period of seven months.

It seems that the Hills case would never have materialised or got as far as it did if it were not for certain magazine articles that came out that did not seem to accurately portray the situation and at that time period of the mid sixties, ridicule was prominent and heaped upon those who claimed such things as conversing with aliens. The Hills wanted the story to be told accurately and handled professionally. Betty was a social worker with a substantial caseload and Barney was a pillar of the community, a member of the human rights activists for racial equality and active in local church groups. In short, they where 'solid' citizens.

Over the years, it has become impossible to refute the possibility that they were in fact abducted; in fact, in the light of further claims of abduction with similar processes experienced by others it was corroborated. The abduction phenomena is an important issue, particularly as it is worldwide, but it is only part of the ET hypothesis.

Since popular culture and the so-called 'ancient astronaut' theories all seem to agree that ET has had a presence on Earth for millennia, the E.B.E. alternatively, extra-terrestrial biological entities of today, even allowing for the 'time dilation' effects when leaving Earth, which they must have done more than once during this lengthy presence, they are most probably the descendants of the original assumed creators that intervened during human development to produce the modern human. The indications are, that this profound event could have occurred 50 thousand years ago and has blended in the ancient folklore and creation stories in all those legends, one of which dare we say may be the Old Testament.

The important point is, as we have stated throughout this work, there must be a 'second coming' with all the startling revelations we must have to accept, which may indeed allow many things that puzzle us to fall neatly into place.

The fact that quite a large number of the human race are Christians will make things easier for many who believe the biblical prophesies regarding the aforesaid second coming. However, it may not be quite how they expect it to be. Certainly the statement "There will be signs in the skies" part of the prophesy, has certainly occurred but the host of angels may well be the descendants of the biblical angels that were active in the days of the Patriarchs that I covered in my book *Pillars of Fire* and that would be a little more disturbing.

An advanced alien race possibly now present in earth space would be well aware of how we would react if they chose to make it perfectly clear and appear 'en-masse', so the aforesaid procedure would be adopted. They must contact the heads of the major powers, Russia, China, the US and all countries that had nuclear weapons stockpiled.

Such leaders would have to be assured that ET 'comes in peace' and suitably convince their military commanders and of course the people and in particular, be alert for any groups that may act on their own volition and ruin the whole programme. There would always be the 'hawks' in the top echelon of the military in all the major powers who would take some firm convincing, if one of the conditions prior to great assistance in technology and all the sciences was to abandon nuclear weapons indeed all weapons and our hard to break habit of killing each other in all the incessant wars. It is staggering to think of all the knowledge they could impart, particularly in the medical field, biology genetics and he ability to cure all the major diseases and afflictions flesh is heir to.

It would be doubtful that they would identify any of the 'hybrids' they may have inserted on Earth after all that embryonic nurturing coupled with decades of abducted females donating their eggs whether they realised it or not. The alleged hybrids may be doing, or have done wonderful work and contributed much to our advancement but once their (to use modern parlance) 'cover had been blown' no one would wish to work alongside them anymore would resent them and despise them as spies and informers. In any case, they would still be useful in their role, perhaps to identify or infiltrate any dangerous resistance groups that may be arising.

Clearly, ET would have much to tell us and we would have many important questions to ask of them. The major one of course would be to explain the mysterious and sudden appearance of modern humans and of course, whether they had been part of any involvement in the process.

We would wish to know if our religious beliefs are all based on a human misunderstanding and how much of the written events many humans have accepted for centuries had been misinterpretations of their involvement.

We could learn much from them in terms of the utilisation of natural resources solar energy, even geothermal energy from beneath the Earth. I must mention a very strange thing, after this last remark regarding 'geothermal energy', I completed that sentence and, since it was around midnight, I retired for the night. The next morning when travelling into the city, I picked up a free copy of the Metro newspaper and read the following: - The deepest and warmest geothermal well has been drilled 3.1 miles down in Penzance Cornwall recently and registered a top temperature of 195°C. The aim is to run a power plant above it, in order to supply electricity to power 3,000 homes, and also, could be used to grow exotic plants and flowers that currently we have to import from abroad.

In one of my books, I mentioned a statement made by a certain Gail Noughton of the Association for the American Advancement of Science that human organs, even the heart would be cellular grown in laboratories. This was before the advent of the amazing 3D printer technology. The aforementioned newspaper I mentioned ran the following article from the *Science Focus* magazine. "The dream of 3D printing whole living human hearts for life saving transplanting operations just got a little closer. A team at Carnegie Mellon University in the US has developed a technique to 3D print Collagen in fine detail, this is a key step to creating replacement organs because collagen, besides being the most abundant protein in the body is a key structural element that forms the biological scaffold that gives organs structure and strength".

The challenge is using collagen, as a 'bio-ink' is that it starts out as a fluid. The technique called 'freeform reversible embedding of suspended hydrogen's (flesh), deposits collagen layer by layer within a support bath of gel. This enables the collagen to solidify in place as the complex 3D structure is built up. When the printing is complete the support gel is melted away by gently heating it to 37°C (body temperature).

If you try to print this in air, it just forms a puddle on your build platform. So we have developed a technique that prevents it from deforming (Andrew Hudson). The technique can print filaments as narrow as -02mm (about the width of a human hair). This enables research to 3D print highly detailed structures into which living cells can be deposited to build muscle and blood vessels. "What we have shown is, that we can print pieces of the heart out of cells and collagen into parts that truly function, like a heart valve, or a small beating ventricle". Amazing stuff.

With regard to the main supposition in this work, that ET 'en-masse' could make their presence finally known to humanity, with the pace of modern advancement in all the sciences, ET would have much less to reveal to us.

However, it cannot be stressed too strongly that a superior intelligence that could traverse the cosmos and perhaps even have been responsible for our very existence, must be enormously far in advance than ourselves, that to attack them in panic would be the most dangerous and stupid thing we could ever do, but alas that is the human way Governments are supposed to employ the services of highly qualified clever scientific minds. We would suggest that they should strive to do their utmost to advise wisely regarding the ET second coming and to ease any forthcoming Cultural Shock humanity would obviously be subjected to.

END

BIBLIOGRAPHY

Edward Ashpole	*The Search for ET Intelligence*	Blandford Press
Isaac Asimov	*Extra Terrestrial Civilisations*	Robson Books 1980
Charles Berlite	*The Rosswell Incident*	Granada 1980
Janet & Colin Bord	*Life Beyond Planet Earth*	Grafton 1991
John G Fuller	*The Interrupted Journey*	Corgi Pubs 1981
Peter Hough & Jenny	*UFOs*	Piatkos 1994
Ian Ridpath	*Messages from the Stars*	Fontana 1978
Carl Sagan	*Cosmos*	Futura 1983
Robert Shapiro	*The Human Blueprint*	Cassel 1992
Stephen Skinner	*Millenium Prophesies*	Carlton 1995
John Spencer	*Gifts of the Gods?*	Virgin 1994
John Spencer	*Perspectives (Abductions)*	McDonald 1990
John Spencer & Hilary Evans	*Phenomenon*	Futura 1988
Jack Stonley	*Is Anyone Out There?*	W.H. Allen 1975